MW00763564

HARCOURT SCIENCE

STANDARDIZED TEST PREPARATION

GRADE 5

TEACHER'S EDITION

Harcourt School Publishers

Orlando • Boston • Dallas • Chicago • San Diego

www.harcourtschool.com

REPRODUCING COPIES FOR STUDENTS

This Teacher's Edition contains full-size student pages with answers printed in nonreproducible blue ink.

It may be necessary to adjust the exposure control on your photocopy machine to a lighter setting to ensure that blue answers do not reproduce.

Copyright © by Harcourt, Inc.

All rights reserved. No part of this publication may be reproduced or transmitted in any form or by any means, electronic or mechanical, including photocopy, recording, or any information storage and retrieval system, without permission in writing from the publisher.

Permission is hereby granted to individual teachers using the corresponding student's textbook or kit as the major vehicle for regular classroom instruction to photocopy Copying Masters from this publication in classroom quantities for instructional use and not for resale. Requests for information on other matters regarding duplication of this work should be addressed to School Permissions and Copyrights, Harcourt, Inc., 6277 Sea Harbor Drive, Orlando, Florida 32887-6777. Fax: 407-345-2418.

HARCOURT and the Harcourt Logo are trademarks of Harcourt, Inc., registered in the United States of America and/or other jurisdictions.

Printed in the United States of America

ISBN 0-15-334080-0

2 3 4 5 6 7 8 9 10 073 09 08 07 06 05 04 03 02

Contents

Unit D — The Solar System and Beyond

Unit E — Building Blocks of Matter

Unit F Energy and Motion

Harcourt Science Standardized Test Preparation
Overview

Using This Resource

The purpose of this practice book is to help you prepare students for success with standardized tests. The book contains practice tests in the same format as many standardized materials. Use of these tests can build students' test-taking skills before the standardized tests are administered. And because the practice test items relate to specific goals and objectives in reading, writing, and mathematics programs, the use of these practice tests can reinforce students' knowledge and skills in the curriculum areas that are assessed.

Description of Components

The *Harcourt Science Standardized Test Preparation* has four parts—reading comprehension, open response, writing practice, and math practice. In each of these areas, the content is drawn from the *Harcourt Science* Pupil Edition (PE). Items in all four areas of the *Test Preparation* are correlated to specific goals and objectives. Following is a brief description of each component.

Daily Reading Comprehension

- A reading comprehension practice test is provided for each lesson in the Pupil Edition (PE) of *Harcourt Science*, Grade 5.

- Students first read a lesson in their textbooks; then they complete related items in their *Standardized Test Preparation* books.

- Each passage is used to test more than one objective.

- Each item includes a purpose statement, a question to answer, or a sentence to complete based on the passage read. Items may also relate to photographs, illustrations, or diagrams.

- Instead of having students mark their answers on the page, you may choose to duplicate the generic answer sheet found in the front of this book.

- Answers are marked on the test pages in nonreproducible blue, so you may copy the test pages and the answers will not duplicate. However, for easy scoring you may wish to make an answer key by using a copy of the answer sheet and filling in the correct answers.

Open Response

- Two short written response questions are provided for each chapter in the Pupil Edition (PE) of *Harcourt Science,* Grade 5.

- Each question in the open response provides a writing prompt from the chapter and includes a hint to help students organize their thoughts.

Writing Practice

- A writing practice exercise is included for each chapter in *Harcourt Science,* Grade 5.

- Practice in writing about science is provided using various forms of discourse—descriptive, expressive, informative, and persuasive.

- Included in the front matter of this booklet are scoring rubrics to help you evaluate each type of writing exercise.

Mathematics Practice

- A math practice test is provided for each chapter in *Harcourt Science,* Grade 5. A math review test is provided for each unit.

- Practice test items deal with chapter science concepts as well as math standards and benchmarks.

- Each item includes a question to answer or a sentence to complete.

- Instead of having students mark their answers on the page, you may choose to duplicate the generic answer sheet found in the front of this book.

- Answers are marked on the test pages in nonreproducible blue, so you may copy the test pages and the answers will not duplicate. However, for easy scoring you may wish to make an answer key by using a copy of the answer sheet and filling in the correct answers.

School-Home Connection

Families often are anxious about state tests and want to help students be successful when taking the tests. You may wish to send home pages from this book with students and encourage family members and students to work together to complete the practice exercises.

Reading Content Standards

The following reading content standards are addressed in the reading comprehension portion of this practice book. Correlations to specific practice pages can be found on the following page.

English Reading: Comprehend a variety of printed materials.

Curriculum Goal

- Recognize, pronounce, and know the meaning of words in text.
    Content Standard: Recognize, pronounce, and know the meaning of words in text by using phonics, language structure, contextual clues, and visual cues.
    Benchmark: Determine meanings of words using contextual clues and illustrations.

Curriculum Goal

- Use a variety of reading strategies to increase comprehension and learning.
    Content Standard: Locate information and clarify meaning by skimming, scanning, close reading, and other reading strategies.
    Benchmark: Locate information using illustrations, tables of contents, glossaries, indexes, headings, graphs, charts, diagrams, and/or tables.

Curriculum Goal

- Demonstrate literal comprehension of a variety of printed materials.
    Content Standard: Demonstrate literal comprehension of a variety of printed materials.
    Benchmark: Identify sequence of events, main ideas, facts, supporting details, and opinions in literary, informative, and practical selections.

Curriculum Goal

- Demonstrate inferential comprehension of a variety of printed materials.
    Content Standard: Demonstrate inferential comprehension of a variety of printed materials.
    Benchmark: Identify relationships, images, patterns, or symbols and draw conclusions about their meanings in printed material.

Curriculum Goal

- Demonstrate evaluative comprehension of a variety of printed materials.
    Content Standard: Demonstrate evaluative comprehension of a variety of printed materials.
    Benchmark: Analyze and evaluate information and form conclusions.

Correlation for Reading Practice

The following reading benchmarks are addressed on the reading comprehension pages of this practice book. In the notations, the letter stands for the related Unit, the first number for the related Chapter, and the second number for the item number in the practice pages.

Reading Benchmarks	Unit-Chapter-Item Number
Benchmark: Determine meanings of words using contextual and structural clues, illustrations, and other reading strategies.	A-1-1, A-1-5, A-1-9, A-1-14, A-1-17, A-2-1, A-2-5, A-2-8, A-2-14, A-2-18, A-3-3, A-3-4, A-3-7, A-3-12, A-3-18, A-4-1, A-4-10, B-1-1, B-1-7, B-1-10, B-2-6, B-2-19, B-3-3, B-3-7, B-4-1, B-4-2, B-4-4, B-4-6, B-4-7, B-4-10, B-4-12, B-4-17, C-1-1, C-2-4, C-2-6, C-2-8, C-2-12, C-2-15, C-3-4, C-3-9, C-3-12, C-3-16, C-4-2, C-4-14, C-4-15, C-4-17, C-4-19, C-4-23, D-1-1, D-1-9, D-1-14, D-2-1, D-2-10, D-2-14, E-1-5, E-1-13, E-1-16, E-2-6, E-2-7, E-2-10, F-1-3, F-1-7, F-1-11, F-1-13, F-2-3, F-2-7, F-3-1, F-3-2, F-3-3, F-3-8, F-3-20, F-3-22, F-4-1, F-4-8, F-4-11
Benchmark: Locate information and clarify meaning by using illustrations, tables of contents, glossaries, indexes, headings, graphs, charts, diagrams, and/or tables.	A-1-2, A-1-6, A-1-8, A-1-11, A-1-15, A-2-2, A-2-6, A-2-15, A-3-1, A-3-11, A-4-8, A-4-13, A-4-16, B-1-3, B-1-6, B-1-11, B-2-2, B-2-5, B-2-11, B-2-14, B-2-23, B-3-1, B-3-6, B-4-11, B-4-20, C-1-3, C-1-7, C-1-13, C-1-15, C-2-2, C-2-7, C-2-9, C-2-13, C-2-14, C-3-1, C-3-6, C-3-7, C-3-8, C-3-10, C-4-5, C-4-8, C-4-10, C-4-18, C-4-21, D-1-2, D-1-8, D-1-10, D-1-12, D-1-18, D-2-4, D-2-8, E-1-6, E-1-10, E-2-2, E-2-5, F-1-6, F-1-17, F-2-4, D-3-10, F-3-12, F-3-13, F-4-2, F-4-13, F-4-14
Benchmark: Identify sequence of events, main ideas, facts, supporting details, and opinions in literary, informative, and practical selections.	A-1-12, A-1-18, A-2-4, A-2-16, A-3-5, A-3-10, A-3-14, A-4-2, A-4-5, A-4-15, A-4-17, B-1-2, B-2-7, B-2-9, B-2-12, B-2-21, B-2-22, B-3-2, B-3-4, B-3-8, B-3-9, B-3-10, B-3-11, B-4-5, B-4-14, B-4-19, B-4-21, C-1-2, C-1-4, C-1-5, C-1-9, C-1-10, C-2-3, C-3-15, C-3-17, C-4-3, C-4-4, C-4-12, C-4-20, C-4-22, D-1-11, D-1-13, D-1-15, D-1-16, D-2-2, D-2-9, D-2-12, D-2-13, D-2-15, D-2-16, D-2-19, E-1-4, E-1-8, E-1-12, E-2-1, E-2-4, E-2-8, E-2-11, E-2-12, F-1-4, F-1-9, F-1-10, F-1-12, F-2-1, F-2-8, F-2-10, F-2-14, F-3-4, F-3-9, F-3-14, F-3-15, F-3-17, F-3-21, F-4-3, F-4-4, F-4-5, F-4-7, F-4-9, F-4-10, F-4-16
Benchmark: Identify relationships, images, patterns, or symbols and draw conclusions about their meanings in printed material.	A-1-3, A-1-10, A-1-13, A-1-16, A-2-10, A-3-2, A-3-13, A-4-4, A-4-11, A-4-14, B-2-3, B-2-10, B-2-17, B-3-12, B-4-13, B-4-16, C-1-8, C-1-12, C-1-14, C-2-18, C-3-13, C-4-9, C-4-11, C-4-13, D-1-3, D-1-5, D-2-3, E-1-2, E-1-3, E-1-7, E-1-14, E-2-3, F-1-2, F-1-5, F-2-2, F-2-5, F-2-15, F-2-16, F-3-5, F-3-6
Benchmark: Analyze and evaluate information and form conclusions.	A-1-4, A-1-7, A-2-3, A-2-7, A-2-9, A-2-11, A-2-12, A-2-13, A-2-17, A-3-6, A-3-8, A-3-9, A-3-15, A-3-16, A-3-17, A-4-3, A-4-6, A-4-7, A-4-9, A-4-12, A-4-18, B-1-4, B-1-5, B-1-8, B-1-9, B-2-1, B-2-4, B-2-8, B-2-13, B-2-15, B-2-16, B-2-18, B-2-20, B-3-5, B-4-3, B-4-8, B-4-9, B-4-15, B-4-18, C-1-6, C-1-11, C-1-16, C-1-17, C-2-1, C-2-5, C-2-11, C-2-16, C-2-17, C-3-2, C-3-3, C-3-5, C-3-11, C-3-16, C-4-1, C-4-6, C-4-7, C-4-16, D-1-4, D-1-6, D-1-7, D-1-17, D-1-19, D-2-5, D-2-6, D-2-7, D-2-11, D-2-17, D-2-18, E-1-1, E-1-9, E-1-11, E-1-15, E-2-9, F-1-1, F-1-8, F-1-14, F-1-15, F-1-16, F-2-6, F-2-9, F-2-11, F-2-12, F-2-13, F-3-7, F-3-11, F-3-16, F-3-18, F-3-19, F-4-6, F-4-12, F-4-15

Writing Content Standards

The following writing content standards are addressed in the open rsponse and writing portions of this practice book. Correlations to specific practice pages can be found on the following page.

Writing: Use writing as a tool to learn, reflect, and communicate for a variety of audiences and purposes.

Curriculum Goal

- Communicate knowledge of the topic, including relevant examples, facts, anecdotes, and details appropriate to topic, audience, and purpose.
    Content Standard: Communicate knowledge of the topic, including relevant examples, facts, anecdotes, and details.
    Benchmark: Convey clear main ideas and supporting details in ways appropriate to topic, audience, and purpose.

Curriculum Goal

- Structure information in clear sequence, making connections and transitions among ideas, sentences, and paragraphs.
    Content Standard: Structure information in clear sequence, making connections and transitions among ideas, paragraphs, and sentences.
    Benchmark: Structure writing by developing a beginning, middle, and end with clear sequencing of ideas and transitions.

Curriculum Goal

- Develop flow and rhythm of sentences.
    Content Standard: Use varied sentence structures and lengths to enhance flow, rhythm, and meaning in writing.
    Benchmark: Use sentence structures that flow and vary in length.

Curriculum Goal

- Demonstrate knowledge of spelling, grammar, punctuation, capitalization, paragraphing, and citing sources.
    Content Standard: Use correct spelling, grammar, punctuation, capitalization, paragraph structure, sentence construction, and other writing conventions.
    Benchmark: Use correct spelling, grammar, punctuation, capitalization, and paragraphing.

Curriculum Goals

- Use a variety of modes (e.g., narrative, imaginative, expository, persuasive) in appropriate context.
- Use a variety of written forms (e.g., journals, essays, short stories, poems, research papers, business and technical writing) to express ideas appropriate to audience and purpose.
- Use multi/step writing process (e.g., identify audience and purpose, generate ideas, plan, draft, confer, revise, and publish) to express ideas.
- Reflect upon and evaluate own writing.
    Content Standard: Use a variety of modes and written forms to express ideas.
    Benchmark: Write in a variety of modes (e.g., narrative, imaginative, expository, persuasive).

Correlation for Writing Practice

The following writing benchmarks are addressed on the Open Response and Writing Practice pages of this book. The notations indicate the pages on which each prompt is found.

Writing Benchmarks	Pages
Benchmark: Convey clear main ideas and supporting details in ways appropriate to topic, audience, and purpose.	Unit A: 4, 5, 9, 10, 14, 15, 19, 20 Unit B: 28, 29, 34, 35, 38, 39, 44, 45 Unit C: 54, 55, 59, 60, 64, 65, 70, 71 Unit D: 80, 81, 85, 86 Unit E: 95, 96, 99, 100 Unit F: 109, 110, 114, 115, 120, 121, 125, 126
Benchmark: Structure writing by developing a beginning, middle, and end with clear sequencing of ideas and transitions.	Unit A: 5, 10, 15, 20 Unit B: 29, 35, 39, 45 Unit C: 55, 60, 64, 71 Unit D: 81, 86 Unit E: 96, 100 Unit F: 110, 115, 121, 126
Benchmark: Use sentence structures that flow and vary in length.	Unit A: 5, 10, 15, 20 Unit B: 29, 35, 39, 45 Unit C: 55, 60, 64, 71 Unit D: 81, 86 Unit E: 96, 100 Unit F: 110, 115, 121, 126
Benchmark: Use correct spelling, grammar, punctuation, capitalization, and paragraphing.	Unit A: 5, 10, 15, 20 Unit B: 29, 35, 39, 45 Unit C: 55, 60, 64, 71 Unit D: 81, 86 Unit E: 96, 100 Unit F: 110, 115, 121, 126
Benchmark: Write in a variety of modes (e.g., narrative, imaginative, expository, persuasive) and forms (e.g., essays, stories, reports) appropriate to audience and purpose.	Unit A: 5, 10, 15, 20 Unit B: 29, 35, 39, 45 Unit C: 55, 60, 64, 71 Unit D: 81, 86 Unit E: 96, 100 Unit F: 110, 115, 121, 126

Mathematics Content Standards

The following mathematics content standards are addressed in the mathematics portion of this practice book. Correlations to specific items on the practice tests can be found on the following page.

CALCULATIONS AND ESTIMATIONS: Select and apply mathematical operations in a variety of contexts.

- COMPUTATION Benchmark: Perform calculations on whole numbers, fractions, and decimals using paper and pencil and calculators.
- NUMBER THEORY Benchmark: Use concepts of primes, factors, and multiples in whole number, fraction, and decimal operations.

MEASUREMENT: Select and use units and tools of measurement.

- UNITS AND TOOLS Benchmark: Select the appropriate units and tools to measure length, perimeter, weight, area, volume, time, temperature, money, and angle.
- DIRECT MEASUREMENT Benchmark: Measure length, perimeter, weight, area, volume, time, temperature, and angle using standard and nonstandard units of measurement.

STATISTICS AND PROBABILITY: Collect, organize, display, interpret, and analyze facts, figures, and other data.

- PROBABILITY Benchmark: Make predictions using experimental probability.
- PROBABILITY Benchmark: Express probabiliaties using fractions, ratios, and decimals.
- INTERPRETATION OF DATA Benchmark: Collect, organize, display, and analyze data using number lines, bar graphs, line graphs, circle graphs, stem and leaf plots, and histograms.

ALGEBRAIC RELATIONSHIPS: Describe and determine generalizations through patterns and functions and represent in multiple ways.

- ESPRESSIONS AND EQUATIONS Benchmark: Use variables and open sentences to express algebraic relationships.
- REPRESENTATIONS OF MATHEMATICAL RELATIONSHIPS Benchmark: Represent and describe relationships among quantities using words, tables, graphs, and rules.

GEOMETRY: Reason about geometric figures and properties and use models, coordinates, and transformational geometry to solve problems.

- CONCEPTS AND PROPERTIES Benchmark: Build, draw, measure, and compare shapes.
- CONCEPTS AND PROPERTIES Benchmark: Visualize and represent two- and three-dimensional geometric figures.

MATHEMATICAL PROBLEM SOLVING: Design, use, and communicate a variety of mathematical strategies to solve problems.

- CONCEPTUAL UNDERSTANDING Benchmark: Use pictures, models, diagrams, and symbols to show main mathematical concepts in the problem.
- CONCEPTUAL UNDERSTANDING Benchmark: Select and use relevant information in the problem to solve it.

Correlation for Math Practice

The following mathematics standards are addressed in this practice book. In the notations, the letter stands for the Unit in the Pupil Edition; the number stands for the item number in the practice test.

Math Standards	Items
Calculations and Estimations: Select and apply mathematical operations in a variety of contexts.	A1, A2, A3, A12, A13, A14, A16, A19, A21, A24, A26, A27, A28, B2, B3, B4, B6, B15, B17, B21, B22, B24, C1, C5, C12, C14, C15, C17, C19, C23, C25, C27, D2, D4, D5, D6, D8, D11, D14, D16, D17, D18, D19, D21, D25, D27, E4, E8, E11, E12, E16, E22, E23, E26, F2, F3, F4, F6, F8, F9, F10, F14, F17, F18, F19, F20, F21, F22, F24, F25, F26, F28, F30
Measurement: Select and use units and tools of measurement.	A4, A10, A11, A20, B16, B20, B23, C2, C3, C4, C8, C11, C16, D2, D3, D4, D5, D6, D10, D11, D12, D13, D18, D20, D23, E13, E15, E17, E20, E21, E28, E30, F9, F10, F11, F12, F16, F21, F22, F27
Statistics and Probability Collect, organize, display, interpret, and analyze facts, figures, and other data.	A5, A6, A7–A9, A15–A17, A18, A22, A30, B1, B2, B5, B6, B8, B9, B10, B11, B12, B13, B14, B19, B26, B28, B29, C6, C9, C13–C15, C18, C22, C24–C26, C28, D1, D7, D22, D24, D28, D29, E1, E5, E6, E7, E14, E18, E19, E25, E27, F1, F5, F9, F13, F23, F29
Algebraic Relationships: Describe and determine generalizations through patterns and functions and represent in multiple ways.	A7, B5, B7, C9, C23, D19, D26, E3, E9, F7, F15, F19, F25, F26, F28
Geometry: Reason about geometric figures and properties and use models, coordinates, and transformational geometry to solve problems.	A23, A29, B7, B18, B25, B27, C7, C20, C29, E2, E10, E29
Mathematical Problem Solving: Design, use, and communicate a variety of mathematical strategies to solve problems.	A25, C10, C14, C15, C21, C23, D9, D15, D19, D21, E3, E24, F3, F4, F15, F19, F25, F26, F28

Rubrics for Writing Practice

A Six-Point Scoring Scale

Student work produced for writing assessment can be scored using a six-point scale. Although the Scoring Guide is comprised of specific descriptors for each score point, each score can also be framed in a more global perspective.

SCORE OF 6: EXEMPLARY. Writing at this level is both exceptional and memorable. It is often characterized by distinctive and unusually sophisticated thought processes with rich details and outstanding craftsmanship.

SCORE OF 5: STRONG. Writing at this level exceeds the standard. It is thorough, complex, and consistently portrays exceptional control of content and skills.

SCORE OF 4: PROFICIENT. Writing at this level meets the standard. It is solid work that has more strengths than weaknesses. The writing demonstrates mastery of skills and reflects considerable care and commitment.

SCORE OF 3: DEVELOPING. Writing at this level shows basic, although sometimes inconsistent mastery and application of content and skills. It shows some strengths but tends to have more weaknesses overall.

SCORE OF 2: EMERGING. Writing at this level is often superficial, fragmented, or incomplete. It may show a partial mastery of content and skills, but it needs considerable development before reflecting the proficient level of performance.

SCORE OF 1: BEGINNING. Writing at this level is minimal. It typically lacks understanding and use of appropriate skills and strategies. Writing at this level may contain major errors.

Rubric for Ideas/Content

Score	Description
6	The writing is exceptionally clear, focused, and interesting. It holds the reader's attention throughout. Main ideas stand out and are developed by strong support and rich details suitable to audience and purpose.
5	The writing is clear, focused, and interesting. It holds the reader's attention. Main ideas stand out and are developed by supporting details suitable to audience and purpose.
4	The writing is clear and focused. The reader can easily understand the main ideas. Support is present, although it may be limited or rather general.
3	The reader can understand the main ideas, although they may be overly broad or simplistic, and the results may not be effective. Supporting detail is often limited, insubstantial, overly general, or occasionally slightly off-topic.
2	Main ideas and purpose are somewhat unclear or development is attempted but minimal.
1	The writing lacks a central idea or purpose.

Rubric for Organization

Score	Description
6	The organization enhances the central idea(s) and its development. The order and structure are compelling and move the reader through the text easily.
5	The organization enhances the central idea(s) and its development. The order and structure are strong and move the reader through the text.
4	Organization is clear and coherent. Order and structure are present, but may seem formulaic.
3	An attempt has been made to organize the writing; however, the overall structure is inconsistent or skeletal.
2	The writing lacks a clear organizational structure. An occasional organizational device is discernible; however, the writing is either difficult to follow and the reader has to reread substantial portions, or the piece is simply too short to demonstrate organizational skills.
1	The writing lacks coherence; organization seems haphazard and disjointed. Even after rereading, the reader remains confused.

Rubric for Sentence Fluency

Score	Description
6	The writing has an effective flow and rhythm. Sentences show a high degree of craftsmanship, with consistently strong and varied structure that makes expressive oral reading easy and enjoyable.
5	The writing has an easy flow and rhythm. Sentences are carefully crafted, with strong and varied structure that makes expressive oral reading easy and enjoyable.
4	The writing flows; however, connections between phrases or sentences may be less than fluid. Sentence patterns are somewhat varied, contributing to ease in oral reading.
3	The writing tends to be mechanical rather than fluid. Occasional awkward constructions may force the reader to slow down or reread.
2	The writing tends to be either choppy or rambling. Awkward constructions often force the reader to slow down or reread.
1	The writing is difficult to follow or to read aloud. Sentences tend to be incomplete, rambling, or very awkward.

Rubric for Word Choice

Score	Description
6	Words convey the intended message in an exceptionally interesting, precise, and natural way appropriate to audience and purpose. The writer employs a rich, broad range of words which have been carefully chosen and thoughtfully placed for impact.
5	Words convey the intended message in an interesting, precise, and natural way appropriate to audience and purpose. The writer employs a broad range of words which have been carefully chosen and thoughtfully placed for impact.
4	Words effectively convey the intended message. The writer employs a variety of words that are functional and appropriate to audience and purpose.
3	Language is quite ordinary, lacking interest, precision, and variety, or may be inappropriate to audience and purpose in places. The writer does not employ a variety of words, producing a sort of "generic" paper filled with familiar words and phrases.
2	Language is monotonous and/or misused, detracting from the meaning and impact.
1	The writing shows an extremely limited vocabulary or is so filled with misuses of words that the meaning is obscured. Only the most general kind of message is communicated because of vague or imprecise language.

Rubric for Conventions

Score	Description
6	The writing demonstrates exceptionally strong control of standard writing conventions (e.g., punctuation, spelling, capitalization, paragraph breaks, grammar, and usage) and uses them effectively to enhance communication. Errors are so few and so minor that the reader can easily skim right over them unless specifically searching for them.
5	The writing demonstrates strong control of standard writing conventions (e.g., punctuation, spelling, capitalization, paragraph breaks, grammar, and usage) and uses them effectively to enhance communication. Errors are so few and so minor that they do not impede readability.
4	The writing demonstrates control of standard writing conventions (e.g., punctuation, spelling, capitalization, paragraph breaks, grammar, and usage). Minor errors, while perhaps noticeable, do not impede readability.
3	The writing demonstrates limited control of standard writing conventions (e.g., punctuation, spelling, capitalization, paragraph breaks, grammar, and usage). Errors begin to impede readability.
2	The writing demonstrates little control of standard writing conventions. Frequent, significant errors impede readability.
1	Numerous errors in usage, spelling, capitalization, and punctuation repeatedly distract the reader and make the text difficult to read. In fact, the severity and frequency of errors are so overwhelming that the reader finds it difficult to focus on the message and must reread for meaning.

Rubric for Voice

Score	Description
6	The writer has chosen a voice appropriate for the topic, purpose, and audience. The writer seems deeply committed to the topic, and there is an exceptional sense of "writing to be read." The writing is expressive, engaging, or sincere.
5	The writer has chosen a voice appropriate for the topic, purpose, and audience. The writer seems committed to the topic, and there is a sense of "writing to be read." The writing is expressive, engaging, or sincere.
4	A voice is present. The writer demonstrates commitment to the topic, and there may be a sense of "writing to be read." In places the writing is expressive, engaging, or sincere.
3	The writer's commitment to the topic seems inconsistent. A sense of the writer may emerge at times; however, the voice is either inappropriately personal or inappropriately impersonal.
2	The writing provides little sense of involvement or commitment. There is no evidence that the writer has chosen a suitable voice.
1	The writing seems to lack a sense of involvement or commitment.

Reading Comprehension—Answer Sheet

Name _____ Date _____

Unit _____ Chapter _____

1. Ⓐ Ⓑ Ⓒ Ⓓ
2. Ⓐ Ⓑ Ⓒ Ⓓ
3. Ⓐ Ⓑ Ⓒ Ⓓ
4. Ⓐ Ⓑ Ⓒ Ⓓ
5. Ⓐ Ⓑ Ⓒ Ⓓ

6. Ⓐ Ⓑ Ⓒ Ⓓ
7. Ⓐ Ⓑ Ⓒ Ⓓ
8. Ⓐ Ⓑ Ⓒ Ⓓ
9. Ⓐ Ⓑ Ⓒ Ⓓ
10. Ⓐ Ⓑ Ⓒ Ⓓ

11. Ⓐ Ⓑ Ⓒ Ⓓ
12. Ⓐ Ⓑ Ⓒ Ⓓ
13. Ⓐ Ⓑ Ⓒ Ⓓ
14. Ⓐ Ⓑ Ⓒ Ⓓ
15. Ⓐ Ⓑ Ⓒ Ⓓ

16. Ⓐ Ⓑ Ⓒ Ⓓ
17. Ⓐ Ⓑ Ⓒ Ⓓ
18. Ⓐ Ⓑ Ⓒ Ⓓ
19. Ⓐ Ⓑ Ⓒ Ⓓ
20. Ⓐ Ⓑ Ⓒ Ⓓ

21. Ⓐ Ⓑ Ⓒ Ⓓ
22. Ⓐ Ⓑ Ⓒ Ⓓ
23. Ⓐ Ⓑ Ⓒ Ⓓ
24. Ⓐ Ⓑ Ⓒ Ⓓ
25. Ⓐ Ⓑ Ⓒ Ⓓ

26. Ⓐ Ⓑ Ⓒ Ⓓ
27. Ⓐ Ⓑ Ⓒ Ⓓ
28. Ⓐ Ⓑ Ⓒ Ⓓ
29. Ⓐ Ⓑ Ⓒ Ⓓ
30. Ⓐ Ⓑ Ⓒ Ⓓ

Math Practice—Answer Sheet

Name _____ Date _____

Unit _____ Chapter _____

1. Ⓐ Ⓑ Ⓒ Ⓓ

2. Ⓐ Ⓑ Ⓒ Ⓓ

3. Ⓐ Ⓑ Ⓒ Ⓓ

4. Ⓐ Ⓑ Ⓒ Ⓓ

5. Ⓐ Ⓑ Ⓒ Ⓓ

6. Ⓐ Ⓑ Ⓒ Ⓓ

7. Ⓐ Ⓑ Ⓒ Ⓓ

8. Ⓐ Ⓑ Ⓒ Ⓓ

9. Ⓐ Ⓑ Ⓒ Ⓓ

10. Ⓐ Ⓑ Ⓒ Ⓓ

11. Ⓐ Ⓑ Ⓒ Ⓓ

12. Ⓐ Ⓑ Ⓒ Ⓓ

13. Ⓐ Ⓑ Ⓒ Ⓓ

14. Ⓐ Ⓑ Ⓒ Ⓓ

15. Ⓐ Ⓑ Ⓒ Ⓓ

16. Ⓐ Ⓑ Ⓒ Ⓓ

17. Ⓐ Ⓑ Ⓒ Ⓓ

18. Ⓐ Ⓑ Ⓒ Ⓓ

19. Ⓐ Ⓑ Ⓒ Ⓓ

20. Ⓐ Ⓑ Ⓒ Ⓓ

21. Ⓐ Ⓑ Ⓒ Ⓓ

22. Ⓐ Ⓑ Ⓒ Ⓓ

23. Ⓐ Ⓑ Ⓒ Ⓓ

24. Ⓐ Ⓑ Ⓒ Ⓓ

25. Ⓐ Ⓑ Ⓒ Ⓓ

26. Ⓐ Ⓑ Ⓒ Ⓓ

27. Ⓐ Ⓑ Ⓒ Ⓓ

28. Ⓐ Ⓑ Ⓒ Ⓓ

29. Ⓐ Ⓑ Ⓒ Ⓓ

30. Ⓐ Ⓑ Ⓒ Ⓓ

TEACHER'S NOTES

What Are Cells, and What Do They Do?

Read pages A6 to A13 in your textbook. Then read each question that follows. Decide which is the best answer to each question. Mark the letter for that answer.

HINT How does the lesson title help you to answer this question?

1. What do scientists call the basic unit of life?

    A a vessel

    B a cell

    C bacteria

    D a nucleus

HINT Reread the first two pages of the lesson to answer the question.

2. Which one is **NOT** a function of every cell in the body?

    A getting rid of cell wastes

    B making new cells for growth and repair

    C carrying signals from the brain to the muscles

    D releasing energy from food

HINT You can check the chart on pages A8 and A9 to find this information quickly.

3. Why might a scientist examine a person's chromosomes?

    A to count how many cells are in the person's body

    B to see if the person's cells are functioning normally

    C to see whether there are bacteria in a cell

    D to get information about a person's characteristics

HINT You can double-check your answer by reading the passage on the same page.

4. What is the purpose of the illustration on page A10?

    A to show how osmosis works

    B to show bacteria in a cell

    C to show how plants get rid of wastes

    D to show how mitochondria produce energy

HINT Use other words in the sentence to help you figure out the meaning of this word.

5. Read this sentence from the lesson.

    The loss of water from the plant's vacuoles causes cytoplasm to shrink away from cell walls.

    What does the word *shrink* mean?

    A get larger

    B move toward

    C expand

    D pull away from

HINT Which heading introduces this information?

6. Which kind of tissue forms the body covering of an animal?

    A connective tissue

    B epithelial tissue

    C nervous tissue

    D muscle tissue

Reading Comprehension

How Do Body Systems Transport Materials?

Read pages A16 to A21 in your textbook. Then read each question that follows. Decide which is the best answer to each question. Mark the letter for that answer.

HINT Think about how the parts of an organism work together.

7. Which of the following famous lines best describes what the section headed "From Cells to Systems" is about?

  A One for all and all for one.

  B Every man for himself.

  C The bigger they are, the harder they fall.

  D Experience is the best teacher.

HINT Look at the diagram on page A19.

8. Where is your liver located?

  A on the left side of your body behind your stomach

  B on the right side of your body in front of your gall bladder

  C on the left side of your body above your small intestine

  D on the right side of your body in front of your stomach

HINT The words *urine, ureter,* and *urethra* all come from the same root word as *urological.*

9. A *urological* disease affects the —

  A digestive system

  B excretory system

  C respiratory system

  D circulatory system

HINT Study the diagrams on pages A18 and A19 to help you answer this question.

10. Which of the following is part of both your digestive and respiratory systems?

  A nose

  B stomach

  C trachea

  D mouth

HINT Skim the lesson looking for information about each organ mentioned in the question.

11. This lesson gives you information on all of the following EXCEPT —

  A how waste products are removed from the body

  B the functions of the blood

  C how the heart and lungs are related

  D how the brain affects appetite

HINT Sequence the events in digestion to answer this question.

12. Food never passes through the —

  A pancreas

  B small intestine

  C esophagus

  D large intestine

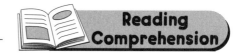

How Do Bones, Muscles, and Nerves Work Together?

Read pages A24 to A27 in your textbook. Then read each question that follows. Decide which is the best answer to each question. Mark the letter for that answer.

HINT Reread page A25 to help you answer this question.

13. What is the difference between tendons and ligaments?

  A Tendons attach muscles to bones. Ligaments are a kind of soft bone.

  B Tendons attach bones to each other. Ligaments attach tendons to each other.

  C Tendons attach bones to muscles. Ligaments attach bones to each other.

  D Tendons attach joints to muscles. Ligaments attach bones to muscles.

HINT Important vocabulary terms sometimes are in italic type.

14. Which is a reflex?

  A blinking

  B chewing

  C running

  D yelling

HINT The illustrations and text on page A26 will help you answer this question.

15. A synapse is —

  A the gap between two neurons

  B a bundle of nerve cells

  C the signal that travels along nerve cells

  D an automatic response to a situation

HINT Which heading introduces this information?

16. The author compares the motion of a ball and socket joint to the motion of a —

  A door hinge

  B wheel

  C joystick

  D on-off switch

HINT Look for words that are similar to cardiologist.

17. A cardiologist is a doctor who specializes in —

  A bones

  B the brain

  C athletic injuries

  D the heart

HINT Clues to the main idea are often found at the beginning of a section.

18. Which sentence states the main idea of the section headed "The Muscular System"?

  A Muscles contract to bend and straighten joints.

  B There are three kinds of muscles — voluntary muscles, smooth muscles, and cardiac muscles.

  C Digestive organs have muscles that help them move substances.

  D The function of cardiac muscles is to move blood.

Unit A, Chapter 1

Base your answers on the information in this chapter. Read all parts to each question before you begin.

The blood transports materials throughout the body. Why does the body need both red blood cells and white blood cells?

HINT What job does each type of blood cell perform?

_____

_____

_____

_____

_____

_____

_____

Your body uses energy to grow and carry out life functions. How do the body's cells get the energy they need?

HINT Reread the passages headed "The Respiratory System" and "The Digestive System" on pages A18 and A19 to find the answer.

_____

_____

_____

_____

_____

_____

Muscle Types

One of the body systems you rely on is your muscular system. This system works with your skeletal and nervous systems so that you can walk, breathe, or move your eyes as you are reading this sentence. In this chapter you learned about different kinds of muscles and what each does in the body. Below are three types of muscles.

voluntary muscles　　smooth muscles　　cardiac muscles

Write an article for your school newspaper explaining about the muscles in the body. Explain what each type does and where each can be found.

Use this page for prewriting or planning activities. Then write your response on a separate sheet of paper.

Writer's Checklist

IDEAS

- Is my message clear?
- Do I know enough about my topic?
- Have I included interesting details?

ORGANIZATION

- Does my paper start out with a bang?
- Did I tell things in the best order?
- At the end does it feel finished and make you think?

VOICE

- Does this writing really sound like me?
- Did I say what I was thinking?
- Did I express how I feel?

WORD CHOICE

- Will my reader understand my words?
- Did I use words I love?
- Are my words interesting?
- Can I picture it?

SENTENCE FLUENCY

- Is my paper easy to read out loud?
- Do my sentences begin in different ways?
- Are some sentences long and some short?

CONVENTIONS

- Did I use paragraphs?
- Is it easy to read my spelling?
- Did I use capital letters in the right place?
- Are periods, commas, exclamation marks, and quotation marks in the right places?

How Do Scientists Classify Living Things?

Read pages A38 to A41 in your textbook. Then read each question that follows. Decide which is the best answer to each question. Mark the letter for that answer.

HINT What did you do when you classified the shoes in the Activity Procedure?

1. What does the word *classify* mean?

   A to name an object

   B to study a nonliving organism

   C to group by using a set of rules

   D to find differences between plants and animals

HINT Often, important information can be found in the first paragraph under a new heading.

2. What is one reason scientists classify living things?

   A to find out how many cells each living thing has

   B to show how a living thing is related to other organisms

   C to explain which type of environment is best for each living thing

   D to perform experiments on these organisms

HINT Living organisms get their nourishment in different ways.

3. Why are fungi not classified as plants?

   A They cannot be eaten by people or animals.

   B They have nuclei.

   C They are monerans.

   D They cannot make their own food.

HINT You can use the text and the chart on page A40 to find this information.

4. What is the smallest classification group used by scientists?

   A species

   B genus

   C order

   D kingdom

HINT Which name do most people recognize: brown bear or *Ursus arctos*?

5. Read these two sentences from the lesson.

   Most living things have a common name such as *brown bear*. But common names may be different in different places.

   What does the word *common* mean?

   A unusual

   B special

   C widely known

   D scientific

HINT Use the chart on page A40 to answer the question.

6. What do scientists call mammals that have sharp teeth for tearing meat?

   A house cats

   B *chordata*, or vertebrates

   C animals

   D *carnivora*, or carnivores

How Are Animals Classified?

Read pages A44 to A47 in your textbook. Then read each question that follows. Decide which is the best answer to each question. Mark the letter for that answer.

HINT Which of the statements below is NOT true?

7. Which is **NOT** a characteristic of animals?

   A   They are made up of many cells.

   B   They eat to survive.

   C   Their cells have nuclei.

   D   They make their own food.

HINT Prefixes can change the meaning of a word.

8. What does the term *invertebrate* mean?

   A   having feathers

   B   not having a backbone

   C   made of many cells

   D   having a backbone

HINT Look at the photographs on pages A44 and A45 to help you answer this question.

9. Which is a true statement about animals?

   A   Snails are a type of vertebrate.

   B   Animals make their own food.

   C   Animals are made of only one cell.

   D   Most vertebrates are larger than invertebrates.

HINT Read the passages under the heading "Animals with a Backbone" to find this information.

10. Amphibians include —

   A   cats and dogs

   B   frogs and toads

   C   owls and robins

   D   sharks and eels

HINT The word *arthropod* means "jointed leg."

11. Which of the following animals are arthropods?

   A   cats and dogs

   B   snails and clams

   C   earthworms and tapeworms

   D   spiders and insects

HINT Read the passage under the heading "A Closer Look at Animals" to help you answer this question.

12. Read these two sentences from the lesson.

   The skeletons of vertebrates are made up of bones that support their bodies from the inside. Most invertebrates have skeletons that form hard outer coverings.

   What can you infer about most animals from this passage?

   A   Most animals have some type of skeleton.

   B   Vertebrates have skeletons, but most invertebrates do not.

   C   Animals do not have skeletons.

   D   Most invertebrates have skeletons, but vertebrates do not.

How Are Plants Classified?

Read pages A50 to A53 in your textbook. Then read each question that follows. Decide which is the best answer to each question. Mark the letter for that answer.

HINT Which of the statements below is NOT true?

13. Which is **NOT** a characteristic of plants?

   A They are made up of many cells.

   B They eat to survive.

   C Their cells have nuclei.

   D They make their own food.

HINT Prefixes can change the meaning of a word.

14. What does the term *nonvascular* mean?

   A having tubes

   B not having tubes

   C made of many cells

   D having roots, stems, and leaves

HINT Under which heading would you find this information?

15. Plants that have tubes are called —

   A nonvascular plants

   B mosses

   C liverworts

   D vascular plants

HINT Look at the diagram on page A51 to help you answer this question.

16. Most of a tree's trunk is made of —

   A bark

   B sapwood

   C heartwood

   D growth rings

HINT Captions as well as the main text can provide important information.

17. You find a plant growing on a rock. This plant is most likely —

   A a moss

   B a fern

   C a liverwort

   D a conifer

HINT Other words in a sentence can help you determine the meaning of a word that you do not know.

18. Read this sentence from the lesson.

   When mosses die, their dead bodies help to enrich the soil, making it more fertile.

   What does the word *fertile* mean?

   A poor

   B having tubes

   C moist

   D productive

Unit A, Chapter 2

Base your answers on the information in this chapter. Read all parts to each question before you begin.

You've learned that there are many types of vertebrates. Write a report for your teacher explaining in what ways invertebrates are alike.

HINT Which type of animals do not have a backbone?

_____

_____

_____

_____

_____

_____

_____

Animals with a backbone are different from those without a backbone. People are one kind of *vertebrate*. What kinds of things does our body allow us to do that invertebrates cannot do? Use information from the lesson along with your own knowledge to write a paragraph for your teacher.

HINT Think about ways you use your body to move and perform activities. Which of these things would be impossible for invertebrates to do?

_____

_____

_____

_____

_____

_____

Classifying Plants

Scientists classify living organisms. For example, animals are classified as vertebrates or invertebrates, depending on whether or not they have a backbone. In a similar manner, plants are classified as vascular or nonvascular, depending on whether or not they contain certain parts.

Write an explanation for a classmate about the important differences between vascular and nonvascular plants. Explain how different types of plants take in water and nutrients.

Use this page for prewriting or planning activities. Then write your response on a separate sheet of paper.

Writer's Checklist

IDEAS

- Is my message clear?
- Do I know enough about my topic?
- Have I included interesting details?

ORGANIZATION

- Does my paper start out with a bang?
- Did I tell things in the best order?
- At the end does it feel finished and make you think?

VOICE

- Does this writing really sound like me?
- Did I say what I was thinking?
- Did I express how I feel?

WORD CHOICE

- Will my reader understand my words?
- Did I use words I love?
- Are my words interesting?
- Can I picture it?

SENTENCE FLUENCY

- Is my paper easy to read out loud?
- Do my sentences begin in different ways?
- Are some sentences long and some short?

CONVENTIONS

- Did I use paragraphs?
- Is it easy to read my spelling?
- Did I use capital letters in the right place?
- Are periods, commas, exclamation marks, and quotation marks in the right places?

Name _____

Date _____

How Do Animals Grow and Reproduce?

Read pages A64 to A69 in your textbook. Then read each question that follows. Decide which is the best answer to each question. Mark the letter for that answer.

HINT Which heading tells about how the body grows?

1. How does the body produce more tissue?

   A It uses tissue from the organs.

   B Cells divide to form new tissue.

   C New cells appear by themselves.

   D The food you eat forms tissue as it is digested.

HINT Reread the information under the heading "Cell Division."

2. What makes each skin cell in a person's body the same?

   A Each cell has the same DNA code.

   B The cells in the body divide by twos.

   C The surface layer of the skin is replaced twice a day.

   D Cells quickly die and reproduce.

HINT Ask yourself how a cell divides. What key words explain this process?

3. What is mitosis?

   A the structure of DNA

   B "threadlike strands"

   C the number of cells in the human body

   D the process of cell division

HINT Use other words in the sentences to help you figure out the meaning of this word.

4. Read these two sentences from the lesson.

   This starfish escaped a hungry predator by dropping one of its arms. When regeneration is complete, the starfish will have a new arm.

   What does the word *regeneration* mean?

   A digestion

   B division of cells

   C tissue replacement

   D injury

HINT Some organisms reproduce asexually.

5. Which is an example of a one-celled organism that reproduces by budding?

   A yeast   **C** starfish

   B planaria   **D** lizards

HINT Reread the last passage of the lesson to find this information.

6. Which statement is TRUE?

   A Every body cell has 92 chromosomes.

   B Gametes have half the number of chromosomes found in body cells.

   C In the second stage of meiosis, cells copy their chromosomes before they divide.

   D Through meiosis, the number of chromosomes in body cells increases.

What Is a Life Cycle?

Read pages A72 to A75 in your textbook. Then read each question that follows. Decide which is the best answer to each question. Mark the letter for that answer.

HINT The word part *meta* means "change."

7. A metamorphosis is a —

   A change in form

   B stage in the life cycle of a animal

   C period of time when no change takes place

   D mistake

HINT Look for the statement that is not supported by information in the lesson.

8. Which is **NOT** a FACT?

   A Cockroaches are born without wings.

   B Many adult insects look completely different from the way they looked when they were born.

   C Only animals that undergo metamorphosis have life cycles.

   D Only animals in the adult stage are able to reproduce.

HINT Compare the description of a caterpillar on page A74 to the pictures on the same page.

9. You can infer from reading this lesson that a caterpillar is —

   A not the same animal as a butterfly

   B a butterfly larva

   C an animal that eats only plants

   D an animal that molts

HINT Reread the section headed "Complete Metamorphosis" to find the answer to this question.

10. Why does all movement stop when an insect enters the pupa stage?

   A The insect doesn't develop legs until it enters the pupa stage.

   B Chemicals prevent the insect from moving.

   C It is during this stage that the insect is laying its eggs.

   D Most of the pupa's energy goes into developing its adult body.

HINT Important information is often found at the beginning of paragraphs.

11. What is the main idea of the section headed "Complete Metamorphosis"?

   A Beetles undergo complete metamorphosis.

   B A larva has no wings and looks very different from its adult form.

   C Egg, larva, pupa, and adult are the four stages of complete metamorphosis.

   D Complete metamorphosis is different from incomplete metamorphosis because it has four stages instead of three.

HINT Captions often include important information.

12. Another word for a *beetle larva* is —

   A pupa   **C** egg

   B adult   **D** grub

Name _____

Date _____

Why Are Offspring Like Their Parents?

Read pages A78 to A81 in your textbook. Then read each question that follows. Decide which is the best answer to each question. Mark the letter for that answer.

HINT The illustration on page A79 will help you answer this question.

13. Two black cocker spaniels have a litter of four puppies. Three of them are black and one of them is buff-colored. You can infer that —

   A the buff color is not an inherited trait

   B buff is a dominant trait

   C buff is a recessive trait

   D the father was buff-colored

HINT Skim the passage to locate important vocabulary in bold type.

14. Recessive traits appear —

   A when the mother has the gene for that trait

   B when both grandparents have the gene for that trait

   C when both parents have the gene for that trait

   D in one out of four offspring

HINT Important words sometimes can be in italic type.

15. This lesson tells you that Mendel crossbred the pea plants. What did he do?

   A He planted only tall green plants.

   B He planted tall green plants next to short yellow plants.

   C He bred tall plants with short plants and yellow-seeded plants with green-seeded plants.

   D He bred pea plants with other kinds of vegetable plants.

HINT Reread page A79 to review Mendel's work.

16. How did Gregor Mendel contribute to our understanding of genetics?

   A He was the first person to do successful gene splicing.

   B His experiments led to the discovery of how traits are inherited.

   C His experiments led to the discovery of chromosomes.

   D He discovered DNA.

HINT Only one of these statements is NOT supported by details from the lesson.

17. Which is **NOT** a FACT?

   A In wild rabbits, dark brown fur is a dominant trait.

   B A dark brown rabbit does not have to have two genes for dark brown fur.

   C Rabbits with dark brown fur are healthier than rabbits with light brown fur.

   D A light brown rabbit must have two genes for light brown fur.

HINT Context clues can help you determine the meaning of a word.

18. In this lesson, the word *hypothesized* means —

   A knew for certain

   B took a wild guess

   C made an educated guess based on observations

   D jumped to a conclusion based on personal feelings and beliefs

Unit A, Chapter 3

Base your answers on the information in this chapter. Read all parts to each question before you begin.

In this chapter you learned that most organisms go through several stages in their life cycles. Explain the type of growth known as direct development.

HINT Remember, different animal species mature in different ways.

_____

_____

_____

_____

_____

_____

_____

The changes in the shape or characteristics of an organism's body as it grows and matures are called metamorphosis. Compare and contrast complete metamorphosis and incomplete metamorphosis.

HINT Use the illustrations on pages A73 and A74 to help you find details to include in your answer.

_____

_____

_____

_____

_____

_____

_____

_____

Explaining Meiosis

Cells are the basic units of all living things. When an organism grows, it grows because the cells that make up the organism have reproduced. Cells reproduce by division. There are two types of cell division: mitosis and meiosis.

Write an article for your school newspaper explaining the importance of meiosis. Briefly describe meiosis and explain how it enables an organism to keep the same number of chromosomes in its body cells after sexual reproduction.

Use this page for prewriting or planning activities. Then write your response on a separate sheet of paper.

Writer's Checklist
IDEAS
• Is my message clear?
• Do I know enough about my topic?
• Have I included interesting details?
ORGANIZATION
• Does my paper start out with a bang?
• Did I tell things in the best order?
• At the end does it feel finished and make you think?
VOICE
• Does this writing really sound like me?
• Did I say what I was thinking?
• Did I express how I feel?
WORD CHOICE
• Will my reader understand my words?
• Did I use words I love?
• Are my words interesting?
• Can I picture it?
SENTENCE FLUENCY
• Is my paper easy to read out loud?
• Do my sentences begin in different ways?
• Are some sentences long and some short?
CONVENTIONS
• Did I use paragraphs?
• Is it easy to read my spelling?
• Did I use capital letters in the right place?
• Are periods, commas, exclamation marks, and quotation marks in the right places?

What Are the Functions of Roots, Stems, and Leaves?

Read pages A92 to A97 in your textbook. Then read each question that follows. Decide which is the best answer to each question. Mark the letter for that answer.

HINT Which other words in the sentence help you to tell the meaning of the word?

1. Read this sentence from the lesson.

   Vascular plants are able to live in different environments because their roots, stems, and leaves are adapted to the environments in which they live.

   Which word below has a meaning almost the same as *adapted?*

   A planted   **C** adjusted

   B grown   **D** connected

HINT Which heading introduces information on roots?

2. What are tree roots that begin above the ground called?

   A prop roots   **C** taproots

   B fibrous roots   **D** root hairs

HINT Reread the passage headed "Stems" to answer the question.

3. Which is **NOT** a function mainly performed by plant stems?

   A Long stems allow plants to store extra water.

   B Stems absorb minerals and water from the soil.

   C The stem allows a plant's leaves to get sunlight even in shady places.

   D Thick stems give some plants extra support.

HINT You can use captions to clarify information.

4. What is the purpose of the illustration on page A94?

   A It shows that a plant's food can be stored in either its stem or its roots.

   B It shows how sugar is processed using plants.

   C It shows that there are different types of plant sugars.

   D It shows that there is more sugar in a beet root than in sugar cane.

HINT What gives plants their green color?

5. Why do the leaves of maple trees turn different colors in autumn?

   A The cold air takes moisture from the leaves.

   B The leaves release different chemicals in cold weather.

   C The leaves of maple trees stop making chlorophyll in the autumn.

   D There is less oxygen in autumn.

HINT What kind of plant has a woody stem?

6. What is the main difference between woody- and soft-stemmed plants?

   A Woody plants live longer than soft-stemmed plants.

   B Woody plants need more sun than soft-stemmed plants.

   C Soft-stemmed plants live longer than woody plants.

   D Woody plants grow in Georgia, unlike soft-stemmed plants.

How Do Plants Reproduce?

Read pages A100 to A107 in your textbook. Then read each question that follows. Decide which is the best answer to each question. Mark the letter for that answer.

HINT Review pages A101 and A102 to answer this question.

7. How are spores different from seeds?

   A Seeds contain a supply of food for growing.

   B Seeds are smaller.

   C A single seed can grow into a new plant. Many spores are needed to produce a new plant.

   D Plants grown from spores cannot produce new plants.

HINT Use the illustrations to help you locate information on pine trees.

8. A pine tree is all of the following EXCEPT —

   A a conifer

   B an angiosperm

   C a gymnosperm

   D a vascular plant

HINT Think about how different plants are pollinated.

9. Which is **NOT** a FACT?

   A In conifers, the male cones produce the pollen.

   B Mosses produce both male and female reproductive cells.

   C The cones of a pine tree are part of the stem.

   D Gymnosperms rely on animals for pollination.

HINT Look for important vocabulary in bold type.

10. What is pollen?

   A structures containing the male reproductive cells of a plant

   B structures containing the female reproductive cells of a plant

   C structures in which seeds develop

   D the seed of a plant

HINT Use the illustration on page A106 to help you answer this question.

11. If you compare a frog's life cycle to the life cycle of a pea plant, the tadpole stage is similar to —

   A the pea pod

   B the seedling

   C the embryo

   D the flowering plant

HINT Review the important ideas covered in this lesson.

12. You could use this lesson to write a report on all of the following EXCEPT —

   A how plants reproduce

   B how plants manufacture food for themselves

   C how to tell the difference between angiosperms and gymnosperms

   D the role seeds play in the life of a plant

Reading Comprehension

How Do People Use Plants?

Read pages A110 to A113 in your textbook. Then read each question that follows. Decide which is the best answer to each question. Mark the letter for that answer.

HINT Use the illustration on page A111 to help you answer this question.

13. Why are fruits and vegetables on the same level of the food guide pyramid?

   A They are equally important.

   B You should eat the same number of servings of fruits and vegetables each day.

   C They have the same amount of fats and calories.

   D They contain the same amounts of vitamins and nutrients.

HINT Read all the captions that surround scientific illustrations.

14. Study the pyramid on page A111. Which of the following is **NOT** a FACT?

   A You should have 3 to 5 servings of vegetables a day.

   B Fish and poultry have more fat than other meats.

   C Foods that are very high in fat are not healthful.

   D Grains are the foundation of a healthful diet.

HINT Compare the information about each food group to answer this question.

15. Which is considered two servings?

   A an apple

   B $\frac{1}{2}$ cup of milk

   C 6 ounces of chopped meat

   D $\frac{1}{2}$ cup chopped spinach

HINT Captions often contain important details.

16. Hand lotion that contains *aloe* has —

   A the outer skin of the aloe plant

   B the crushed flowers of the aloe plant

   C the leaves of the aloe plant

   D stored food of the aloe plant

HINT Important information is often found at the beginning of paragraphs.

17. What is the main idea of the section headed "Plants as Medicines"?

   A Many medicines we use today were discovered by Native Americans.

   B Almost half of all the medicines we use today are made from plants.

   C Most medicines come from the bark of trees.

   D Aspirin is almost identical to a substance found in the bark of the willow tree.

HINT Review the lesson for important details.

18. All of these statements are general-izations EXCEPT —

   A some medicines are found in the bark of trees

   B foods in the Fats, Oils, and Sweets group are generally unhealthful

   C homes are often made of wood, and a lot of furniture in most homes is wood

   D it takes about 100 kg of rose petals to make 30 mL of fragrance

Unit A, Chapter 4

Base your answers on the information in this chapter. Read all parts to each question before you begin.

You've learned that it takes a variety of foods to keep your body healthy. Use the Food Guide Pyramid on page A111 and plan a healthy menu for one day. Your menu should include three meals and several small snacks. You can use examples from the pyramid as well as your own food choices in your menu.

HINT The captions on page A11 explain the number of servings recommended for each type of food.

_____

_____

_____

_____

_____

_____

_____

_____

You've learned that people use plants in many ways. Explain how you use plants every day. Include how many times you use them and for what purposes. Include your ideas about whether it would be hard to do without plants.

HINT Review the photographs in Lesson 3 to give you ideas as to ways you use plants every day.

_____

_____

_____

_____

_____

_____

Plant Reproduction

Most of the plants you are familiar with are probably angiosperms, or flowering plants. You learned in this chapter that gymnosperms, or plants with cones, depend on the wind to get pollen from the male cone to the female cone. Although angiosperms can be pollinated by the wind, they also depend on other methods.

Write a report for your teacher that explains how pollination takes place in flowering plants. Describe at least two methods.

Use this page for prewriting or planning activities. Then write your response on a separate sheet of paper.

Writer's Checklist	
IDEAS • Is my message clear? • Do I know enough about my topic? • Have I included interesting details?	**WORD CHOICE** • Will my reader understand my words? • Did I use words I love? • Are my words interesting? • Can I picture it?
ORGANIZATION • Does my paper start out with a bang? • Did I tell things in the best order? • At the end does it feel finished and make you think?	**SENTENCE FLUENCY** • Is my paper easy to read out loud? • Do my sentences begin in different ways? • Are some sentences long and some short?
VOICE • Does this writing really sound like me? • Did I say what I was thinking? • Did I express how I feel?	**CONVENTIONS** • Did I use paragraphs? • Is it easy to read my spelling? • Did I use capital letters in the right place? • Are periods, commas, exclamation marks, and quotation marks in the right places?

Unit A, Chapter 1

Read each question and choose the best answer. Mark the letter for that answer.

1. A plant cell has 10 molecules of water. After the plant is watered, the molecules of water in the plant cell increase to 22 molecules of water through osmosis. How many more molecules of water does the plant cell have after osmosis?

   A 12 molecules

   B 14 molecules

   C 20 molecules

   D 22 molecules

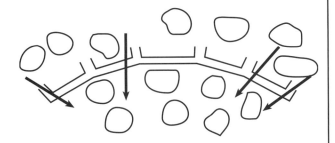

2. The above picture shows water molecules that will move through a plant cell wall by osmosis. How many water molecules will be in the plant cell after osmosis?

   A 0 molecules

   B 7 molecules

   C 14 molecules

   D cannot be determined

3. Juanita's heart beats 20 times in 15 seconds. How many times will it beat in one minute?

   A 15

   B 80

   C 300

   D 1200

4. If Jose gains 3000 milliliters in fluids in a day, how many liters did he gain?

   A 0.03 L

   B 0.3 L

   C 3 L

   D 30 L

Refer to the table below for problems 5 and 6.

Distance Around Upper Arm

Name	Straight Arm	Bent Arm
Randy	5 in.	7 in.
William	4 in.	$5\frac{1}{2}$ in.
Eleanor	$6\frac{1}{2}$ in.	$8\frac{1}{4}$ in.
Anthony	$4\frac{1}{2}$ in.	$6\frac{1}{4}$ in.

5. Which person showed the least increase in distance around upper arm after the arm is bent?

   A Randy

   B William

   C Eleanor

   D Anthony

6. How many inches did most of the students' arms increase when their arms went from straight to bent?

   A $1\frac{3}{4}$ in.

   B $1\frac{7}{8}$ in.

   C 2 in.

   D $2\frac{1}{2}$ in.

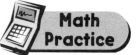

Unit A, Chapter 2

Read each question and choose the best answer. Mark the letter for that answer.

Use the bar graph for problems 7 to 9.

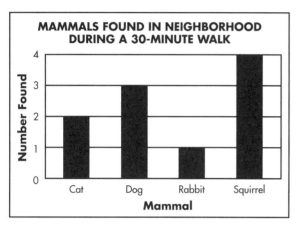

7. Determine how many dogs were seen.

A 1   **C** 3

B 2   **D** 4

8. Which type of mammal are you most likely to see while walking in that neighborhood?

A cat   **C** rabbit

B dog   **D** squirrel

9. Jake went for a walk under the same conditions as the data in the graph above were collected under. He said he saw 100 rabbits and 1 squirrel. Is this a reasonable statement?

A No, it is not reasonable. Jake should have seen more squirrels than rabbits.

B Yes, it is reasonable.

C No, it is not reasonable. Jake should have seen more than 100 rabbits.

D No, it is not reasonable. Jake should have seen at least 3 squirrels.

10. If you are asked to observe an experiment from 10:43 A.M. to 11:02 A.M., for how many minutes do you have to watch?

A 15 minutes

B 19 minutes

C 20 minutes

D 60 minutes

11. Which is the most reasonable measurement of the amount of liquid in the dropper below?

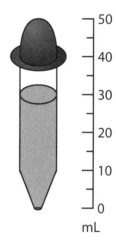

A 1 mL   **C** 40 mL

B 30 mL   **D** 100 mL

12. There are 750,000 species of insects. There are 20,000 species of orchids. How many more species of insects are there than orchids?

A 770,000

B 752,000

C 730,000

D 950,000

Unit A, Chapter 3

Read each question and choose the best answer. Mark the letter for that answer.

13. An alligator has 32 chromosomes in its body cells. Determine the number of chromosomes in an alligator's reproductive cells.

   A 8 chromosomes

   B 16 chromosomes

   C 30 chromosomes

   D 64 chromosomes

14. A pigeon has 80 chromosomes in its body cells. Determine the number of chromosomes in a pigeon's zygotes.

   A 20 chromosomes

   B 40 chromosomes

   C 80 chromosomes

   D 160 chromosomes

Use the bar graph below for problems 15 to 17.

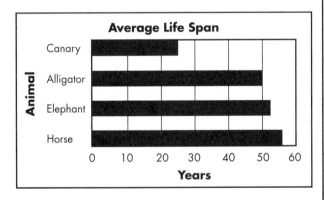

15. Which animal has the shortest life span?

   A horse   **C** alligator

   B elephant   **D** canary

16. About how many more years does a horse live than a canary?

   A 20 years   **C** 30 years

   B 25 years   **D** 35 years

17. Which of the following statements is reasonable?

   A An elephant will live longer than a canary.

   B A canary will live longer than a horse.

   C An alligator will live longer than an elephant.

   D A canary lives longer than most animals.

18. Which of the following bar graphs represents the data in the table below?

Eye Color Survey

Eye Color	Number of Students
Green	6
Brown	12
Blue	9

A
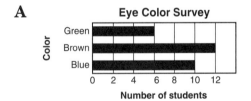

B
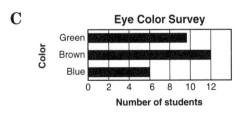

C
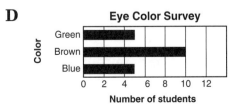

D

Unit A, Chapter 4

Read each question and choose the best answer. Mark the letter for that answer.

19. The length of the root of a pepper plant is 5.9 centimeters long. The length of the root of a tomato plant is 9.2 centimeters long. How much longer is the tomato plant root than the pepper plant root?

  A 3.3 cm   **C** 4.7 cm

  B 4 cm   **D** 5 cm

20. The length of a leaf is 10 millimeters. How many centimeters long is the leaf?

  A 1 cm   **C** 100 cm

  B 10 cm   **D** 1000 cm

21. Female cones can be as short as 2 centimeters and as long as 750 centimeters. What is the difference in centimeters from the largest female cone to the smallest female cone?

  A 742 cm   **C** 750 cm

  B 748 cm   **D** 752 cm

22. According to the circle graph below, which of the following are you more likely to see while walking in the park?

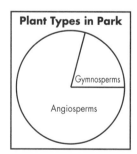

  A angiosperms

  B gymnosperms

  C Both are equally likely.

  D cannot be determined

23. If this box is filled with popcorn, what is the volume of popcorn in the box?

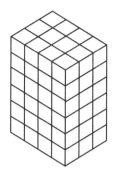

  A 10 cubic units

  B 13 cubic units

  C 30 cubic units

  D 72 cubic units

24. Using the table below, determine how many grams of fat Kayla consumed during dinner.

Kayla's Dinner

Item	Fat Grams
Hamburger	60 g
Fries	80 g
Soda	0 g

  A 60 g   **C** 140 g

  B 80 g   **D** 1400 g

Unit A, Review

Read each question and choose the best answer. Mark the letter for that answer.

25. As blood travels through a body it gains carbon dioxide through diffusion. How will the amount of carbon dioxide change from the time the blood leaves the lungs to the end of its travels?

A increase

B decrease

C stay the same

D cannot be determined

26. According to the table below, how many beats per minute slower is Charles' heartbeat rate than Scott's?

Heart Rate Survey

Person	Beats per Minute
JoAnn	80
Scott	79
Chemika	63
Charles	71

A 1   **C** 6

B 5   **D** 8

27. A pigeon has 80 chromosomes in its body cells. Determine how many chromosomes are in the pigeon's reproductive cells.

A 30 chromosomes

B 40 chromosomes

C 80 chromosomes

D 160 chromosomes

28. The length of a squash plant root is 15.7 centimeters. The length of a lima bean plant root is 8.1 centimeters. How much longer is the squash plant root than the lima bean plant root?

A 2.6 cm   **C** 7.6 cm

B 5.8 cm   **D** 8.6 cm

29. What is the volume of the box of peas below?

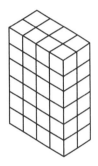

A 8 cubic units   **C** 40 cubic units

B 12 cubic units   **D** 48 cubic units

30. According to the circle graph below, which is more likely to be found in the garden?

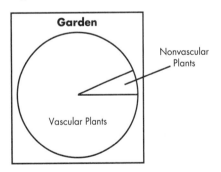

A vascular plants

B nonvascular plants

C neither vascular nor nonvascular plants

D not enough information to tell

How Does Nature Reuse Materials?

Read pages B6 to B11 in your textbook. Then read each question that follows. Decide which is the best answer to each question. Mark the letter for that answer.

HINT Context clues can help you determine the meaning of unknown words.

1. In this lesson, the word *fixed* means —

   A changed   **C** recycled

   B set   **D** repaired

HINT Under which heading would you find this information?

2. How do animals get most of the nitrogen they need to live?

   A from the air by breathing

   B from the water

   C by eating plants and plant-eating animals

   D by changing the oxygen they breathe into nitrogen

HINT Skim the passage on page B10 to help you locate this information.

3. What effect did the Industrial Revolution have on the carbon-oxygen cycle?

   A More carbon dioxide was added to the air through the burning of fossil fuels.

   B Less carbon dioxide was available because more crops were grown to feed an increasing population.

   C As the population increased, more and more carbon dioxide was added to the air.

   D As cities sprang up, forests were cut down. With fewer trees, the amount of oxygen in the atmosphere fell dramatically.

HINT Use the diagram on page B7 to help you summarize the nitrogen cycle.

4. Which of the following best summarizes the nitrogen cycle?

   A Plants and animals remove nitrogen from Earth and replace it with their waste and when they die. Lightning and bacteria change atmospheric nitrogen into a form that plants and animals can use.

   B Animals eat plants that contain nitrogen. They return the nitrogen to the soil through their wastes.

   C Bacteria and lightning change the nitrogen in the soil into a form that plants and animals can use. Plants absorb the nitrogen through their roots, and then the animals eat the plants.

   D Farmers spread animal wastes on their soil to return some of the lost nitrogen and increase their crop yields.

HINT Reread the captions on pages B8 and B9 to help you answer this question.

5. Which statement is **NOT** correct?

   A All animals use oxygen and release carbon dioxide.

   B Plants use carbon dioxide for photosynthesis and release oxygen.

   C Microscopic plantlike organisms in the oceans release oxygen into the water.

   D During respiration, plants release oxygen into the air.

Why Is the Water Cycle Important?

Read pages B14 to B17 in your textbook. Then read each question that follows. Decide which is the best answer to each question. Mark the letter for that answer.

HINT The diagram on page B14 will help you answer this.

6. Earth has —

   A almost equal amounts of fresh and salt water

   B mostly salt water

   C more fresh water in the ground than in the air

   D almost no fresh water in the ice caps

HINT Important words are often in bold type.

7. What is transpiration?

   A rain, snow, and other forms of water that fall from clouds

   B the movement of water through the water cycle

   C the process by which plants give off water through their stomata

   D the process by which water is returned to Earth

HINT Think about the temperature changes that help drive the water cycle.

8. Condensation occurs when —

   A water is warmed and becomes water vapor

   B water freezes and becomes ice

   C water vapor is cooled and becomes liquid water

   D ice is warmed and becomes liquid water

HINT Under what heading would you find this information?

9. After reading this lesson you can draw all of the following conclusions EXCEPT —

   A the amount of water on Earth will never change

   B as long as the water cycle continues, the water on Earth will always be fresh

   C the sun is the driving force behind the water cycle

   D there are ways to conserve water

HINT Use context clues to help you determine the meaning of unfamiliar words.

10. A synonym for *arid* is —

   A dry   **C** hot

   B dusty   **D** watery

HINT Clues in the surrounding sentences can help you understand unknown words.

11. What is groundwater?

   A water from a heavy rain that soaks into the ground

   B water that falls and collects in puddles

   C water that collects underground and supplies wells and springs

   D the water trapped in the ice caps

Unit B, Chapter 1

Base your answers on the information in this chapter. Read all parts to each question before you begin.

You've learned that many materials are used and then reused by living organisms. Nitrogen, carbon, oxygen, and water are some of the materials recycled by nature. Describe the nitrogen cycle. Include both living and nonliving parts of the cycle.

HINT The illustration and captions on page B7 will help you answer this question.

Possible answer: Bacteria, and sometimes lightning, change nitrogen gas into nitrates and ammonia. Plants use these fixed forms of nitrogen to make proteins. Animals get the nitrogen they need by eating plants and other animals. Animal wastes and decaying organisms return usable nitrogen to the soil for more plants to use.

Until about 200 years ago, the carbon–oxygen cycle was balanced. But in the past 200 years, several things have changed the balance of this cycle. Describe at least two things that have helped to change this balance, and why this change is a problem.

HINT Include the role of people in your response.

Possible answer: Human activity since the beginning of the Industrial Revolution has caused changes in the carbon–oxygen cycle. The increased burning of wood and coal has put tons of carbon dioxide into the air every year. The replacement of wood and coal by fuels such as natural gas and petroleum adds even more carbon dioxide to the air. Many forests have been cut down to provide fuel for factories and to provide room for other human needs. With fewer forests, the rate of removal of carbon dioxide from the atmosphere has decreased. This build-up of carbon dioxide is a problem, because excess carbon dioxide is poisonous to animals, including humans.

Water Quality

In this chapter you learned that about 97 percent of the water on Earth is salt water. Most of the fresh water is locked up in glaciers and ice caps. As a result, very little of Earth's fresh water is available for human use. Because this water is so precious, the United States government has passed laws to preserve and improve the quality of the nation's water.

Find out about the Clean Water Act or other laws that protect surface water and ground water. Write a letter to the Environmental Protection Agency or to your representative in Congress. In your letter, explain why you think water quality is important. Then ask for more information about the laws that protect our water.

Use this page for prewriting or planning activities. Then write your response on a separate sheet of paper.

Writer's Checklist

IDEAS

- Is my message clear?
- Do I know enough about my topic?
- Have I included interesting details?

ORGANIZATION

- Does my paper start out with a bang?
- Did I tell things in the best order?
- At the end does it feel finished and make you think?

VOICE

- Does this writing really sound like me?
- Did I say what I was thinking?
- Did I express how I feel?

WORD CHOICE

- Will my reader understand my words?
- Did I use words I love?
- Are my words interesting?
- Can I picture it?

SENTENCE FLUENCY

- Is my paper easy to read out loud?
- Do my sentences begin in different ways?
- Are some sentences long and some short?

CONVENTIONS

- Did I use paragraphs?
- Is it easy to read my spelling?
- Did I use capital letters in the right place?
- Are periods, commas, exclamation marks, and quotation marks in the right places?

What Are Ecosystems?

Read pages B28 to B31 in your textbook. Then read each question that follows. Decide which is the best answer to each question. Mark the letter for that answer.

HINT What is a nonliving thing?

1. Which is **NOT** part of the physical environment?

   A landforms

   B air

   C animals

   D water

HINT Look under the first heading of the lesson to find this information.

2. What are populations of organisms living together called?

   A an organism

   B a community

   C an ecosystem

   D a habitat

HINT How would you compare a habitat to a niche?

3. How is the niche of a great horned owl different from that of a golden eagle?

   A The owl and the eagle have different habitats.

   B The owl and the eagle eat different prey.

   C The owl and the eagle live in different environments.

   D The owl hunts at night, and the eagle hunts in the day.

HINT Are populations made up of living or nonliving things?

4. Which is an example of how populations depend upon one another?

   A Great horned owls eat mice.

   B Grasshoppers reproduce by mating with other grasshoppers.

   C Water can completely cover a landform.

   D Carbon dioxide and oxygen remain in balance in a healthy environment.

HINT Use the charts on page B30 to help you answer this question.

5. How much more precipitation falls in January on Southern Vermont than on Southern Arizona?

   A less than 20 inches more

   B over 25 inches more

   C about 22 inches more

   D almost 19 inches more

HINT Use context clues to find the meaning of the word.

6. Read this sentence from the lesson.

   The amount of food—or any limited resource—in an ecosystem affects the size of a population.

   Which word means almost the opposite of the word *limited?*

   A small   C nonliving

   B changing   D endless

How Does Energy Flow Through an Ecosystem?

Read pages B34 to B39 in your textbook. Then read each question that follows. Decide which is the best answer to each question. Mark the letter for that answer.

HINT Read all captions in scientific illustrations.

7. The food web on pages B36 and B37 shows —

   A a western harvest mouse feeding on white prairie clover

   B a bison feeding on side-oats grama grass

   C a thirteen-lined ground squirrel feeding in the buffalo grass

   D a two-striped grasshopper beside the purple coneflower

HINT Review pages B34 and B35 to help you answer this question.

8. Which of the following is **NOT** correctly paired?

   A bacteria—decomposer

   B zebra—consumer

   C clover—decomposer

   D maple tree—producer

HINT Think about the roles of different organisms in food chains.

9. Why will you often see mushrooms growing on a rotting log?

   A Mushrooms can only grow in soft wood.

   B Mushrooms are decomposers. They break down dead tissue and use the nutrients for growth.

   C Mushrooms are consumers, and as the log rots they are able to get the nourishment they need.

   D Mushrooms feed on the insects that attack the log.

HINT Where are hawks located in the pyramid?

10. In the energy pyramid, hawks are —

    A producers

    B first level consumers

    C second level consumers

    D third level consumers

HINT Important vocabulary often is indicated by bold type.

11. An energy pyramid shows —

    A the amount of energy used by each consumer in the food chain

    B the amount of energy available to each level of a food chain

    C how sunlight affects the producers in a food chain

    D how members of a food chain interact with each other

HINT You often can find important information at the beginning of a section.

12. Which of the following states the main idea of the section headed "Food Chains"?

    A A food chain shows how the consumers in an ecosystem are connected to one another according to what they eat.

    B Identifying the organisms and their levels can help you understand how energy moves through an ecosystem.

    C A food chain has several levels.

    D At the base of the food chain are the plants, and consumers make up the other levels.

How Do Organisms Compete and Survive in an Ecosystem?

Read pages B42 to B47 in your textbook. Then read each question that follows. Decide which is the best answer to each question. Mark the letter for that answer.

HINT Important vocabulary sometimes appears in italic type.

13. Which animal is using camouflage to escape its enemies?

   A a deer, who relies on speed

   B a porcupine, who curls into a sharp-spined ball

   C a skunk who sprays a foul-smelling, stinging liquid

   D a moth, whose wing colors blend in with the bark of a tree

HINT How are these two insects interacting?

14. The ants and aphids shown in the picture on page B45 illustrate —

   A competition

   B mutualism

   C predator and prey

   D symbiosis where only one organism benefits

HINT Under which heading would you look for this information?

15. Which of the following is an instinct?

   A ground squirrels hibernating during the winter

   B dolphins catching rings on their snouts

   C parrots asking for crackers

   D a house cat rushing into the kitchen when it hears the sound of a can opener

HINT You must read both the text and the captions on pages B43 and B44 to answer this question.

16. How is the relationship among giraffes, antelopes, and rhinos similar to the relationship among blackburnian, bay-breasted, and myrtle warblers?

   A Both groups are in fierce competition for food.

   B In both groups, one member is a carnivore that preys on the other members.

   C In both groups, the animals have a symbiotic relationship with each other.

   D The animals in both groups do not compete with each other because they feed from different tree levels.

HINT Use your knowledge of animal behavior to help you answer this question.

17. All of these animals are displaying learned behavior EXCEPT —

   A cats purring

   B raccoons looking for food in garbage cans

   C an owl catching a mouse

   D osprey building their nests on top of telephone poles instead of in trees

Name _____

Date _____

What Is Extinction and What Are Its Causes?

Read pages B50 to B53 in your textbook. Then read each question that follows. Decide which is the best answer to each question. Mark the letter for that answer.

HINT Which heading in the lesson introduces this topic?

18. What is the main cause of a decline in population numbers?

   A natural disasters, such as floods

   B diseases

   C human activity

   D cold temperatures

HINT Use the definitions following the key words highlighted in yellow to answer the question.

19. What do you call a population when the last of its members dies?

   A declining   **C** threatened

   B extinct   **D** endangered

HINT Although DDT had more than one harmful effect on these birds, the passage states that one result was especially dangerous to the bald eagle population.

20. How did the chemical pesticide DDT directly harm bald eagles?

   A DDT poisoned the adult birds, which then died.

   B DDT killed the trees in which the bald eagles nested.

   C DDT weakened the eggshells of the bald eagles' offspring.

   D DDT poisoned the rivers and streams in the bald eagles' habitats.

HINT "Rate" means how fast something happens.

21. Over the past two hundred years, what has happened to the rate of extinction?

   A It has slowed down.

   B It has remained the same.

   C It has speeded up 1000 times.

   D It has speeded up only slightly.

HINT Reread the passage following the heading "Extinction Is Forever" to find this information.

22. How many animals must remain in a population in order for that population to have a good chance of survival?

   A 50   **C** 1000

   B 100   **D** over 1000

HINT Headings give clues about the information contained in a passage.

23. What populations would you expect to learn about from the passage headed "Success Stories"?

   A animals recently placed on the endangered species list

   B animals who have become extinct over the past 200 years

   C animals who have large, thriving populations

   D animals whose populations have been saved from extinction

Name _____

Date _____

Unit B, Chapter 2

Base your answers on the information in this chapter. Read all parts to each question before you begin.

Food chains show how the organisms in an ecosystem are connected to one another according to what they eat. Every food chain has several levels. Describe the levels of a food chain.

HINT What do organisms eat at each level of a food chain?

_____

_____

_____

_____

_____

_____

_____

_____

An energy pyramid shows the amount of energy available to pass from one level of a food chain to the next. Explain why less energy is available to organisms at the top of the energy pyramid. Include how this affects populations of animals such as wolves.

HINT Review the text and diagram on page B38.

_____

_____

_____

_____

_____

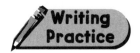
Producers and Consumers

An energy pyramid begins with producers, organisms that make their own food. The next level in the pyramid is occupied by consumers, which eat the producers, and are, in turn, eaten by other consumers at higher levels of the pyramid. Recall that as you pass upward through the pyramid, less energy is available to the organisms on the higher levels.

It is the opinion of some scientists that if humans ate more producers and fewer consumers, more food could be produced for less money. Write a letter to the editor of your local newspaper in which you support their opinion with facts you learned in this chapter.

Use this page for prewriting or planning activities. Then write your response on a separate sheet of paper.

Writer's Checklist

IDEAS
- Is my message clear?
- Do I know enough about my topic?
- Have I included interesting details?

ORGANIZATION
- Does my paper start out with a bang?
- Did I tell things in the best order?
- At the end does it feel finished and make you think?

VOICE
- Does this writing really sound like me?
- Did I say what I was thinking?
- Did I express how I feel?

WORD CHOICE
- Will my reader understand my words?
- Did I use words I love?
- Are my words interesting?
- Can I picture it?

SENTENCE FLUENCY
- Is my paper easy to read out loud?
- Do my sentences begin in different ways?
- Are some sentences long and some short?

CONVENTIONS
- Did I use paragraphs?
- Is it easy to read my spelling?
- Did I use capital letters in the right place?
- Are periods, commas, exclamation marks, and quotation marks in the right places?

What Are Land Biomes?

Read pages B64 to B73 in your textbook. Then read each question that follows. Decide which is the best answer to each question. Mark the letter for that answer.

HINT You can locate this information in the table on page B72.

1. The taiga receives —

   A less rainfall than the desert

   B more rainfall than a deciduous forest

   C more rainfall than the grasslands

   D more rainfall than tropical rain forests

HINT Under which heading would you find this information?

2. Why are North American grasslands called the world's "breadbasket"?

   A Most of the world's bread is produced in America.

   B More wheat is grown in the United States than anywhere else.

   C The art of breadmaking was first developed by the colonists.

   D The North American grasslands are used to produce tremendous amounts of food.

HINT Summarize the section titled "Tundra."

3. Which words best describe a tundra?

   A long, dark winters and permafrost

   B cold winters and very hot summers with little rain

   C permanent ice caps and heavy snowfall year round

   D drought, very few plants, and 14 hours of sunlight year round

HINT Skim the passage about deserts to locate this information.

4. How is mesquite different from most other desert plants?

   A It cannot store water; instead, its roots grow deep to reach underground water.

   B It is poorly adapted to a hot, dry climate and is just barely able to survive.

   C Its leaves form a canopy over the plant to shade it from the hot sun.

   D It is the only plant in the world that is able to survive for years without water.

HINT Think about how evergreens and deciduous plants differ.

5. Evergreens in the taiga can make food year-round because —

   A they don't shed their leaves all at once

   B their needles have a waxy coating

   C the climate in a taiga never changes

   D when the needles drop off, the bark takes over the food-making

HINT Study the map on pages B64 and B65 to help you answer this question.

6. There are no deciduous forests in —

   A South America

   B Africa

   C North America

   D Australia

What Are Water Ecosystems?

Read pages B76 to B81 in your textbook. Then read each question that follows. Decide which is the best answer to each question. Mark the letter for that answer.

HINT Important words sometimes appear in italic type.

7. A brackish-water ecosystem is one in which —

  A there are many pollutants

  B the water is above 70°F

  C salt and fresh water mix

  D there are few currents and the water barely moves

HINT Think about how plants make food.

8. Why aren't plants able to survive below a depth of 200 meters?

  A It is too cold.

  B The plants cannot anchor themselves in the shifting sands.

  C There is no sunlight at that depth.

  D The water pressure is too great.

HINT Carnivores eat other animals.

9. Which is in the right order?

  A Phytoplankton eat zooplankton, which then eat small fish. Large carnivores eat the small fish.

  B Zooplankton eat phytoplankton. The zooplankton are eaten by small fish, which are in turn eaten by larger carnivores.

  C Large carnivores eat small fish. Small fish eat phytoplankton. Phytoplankton eat zooplankton.

  D Zooplankton eat phytoplankton and small fish. Large carnivores eat phytoplankton, zooplankton, and small fish.

HINT Read carefully to locate details in a reading passage.

10. What is a bluegill?

  A a pond plant

  B a small fish

  C a type of water fowl

  D an insect

HINT Under which heading would you find this information?

11. The water in an estuary —

  A changes from salt to fresh as ocean tides rise and fall

  B contains few nutrients

  C is generally too changeable to support very many animals

  D is often several hundred meters deep

HINT Examine the illustration on pages B76 and B77 to help you answer this question.

12. This is what Captain Jack caught. Where was he fishing?

  A in the intertidal zone

  B in the open ocean zone

  C in an estuary

  D near-shore zone

Open Response

Unit B, Chapter 3

Base your answers on the information in this chapter. Read all parts to each question before you begin.

A biome is a large-scale ecosystem. Earth has six major biomes. Use information from the text and your own observations and experiences to write a description of the biome in which you live.

HINT Identify your biome using the map on pages B64 and B65.

_____

_____

_____

_____

You have learned that there are several types of forest biomes. Compare and contrast the tropical rain forest biome with the deciduous forest biome.

HINT How does climate affect vegetation?

_____

_____

_____

_____

_____

_____

_____

_____

_____

_____

Name _____

Date _____

Saltwater Ecosystems

The ecosystems you are most familiar with are probably land biomes. However, the ocean has a number of habitats for organisms as well. In this chapter you learned about the following three types of salt water ecosystems.

intertidal zone   near-shore zone   open-ocean zone

Pick one of the ecosystems listed above. Write a story for a younger student that includes a description of that ecosystem. Include examples of the types of organisms found in that ecosystem.

Use this page for prewriting or planning activities. Then write your response on a separate sheet of paper.

Writer's Checklist

IDEAS
- Is my message clear?
- Do I know enough about my topic?
- Have I included interesting details?

ORGANIZATION
- Does my paper start out with a bang?
- Did I tell things in the best order?
- At the end does it feel finished and make you think?

VOICE
- Does this writing really sound like me?
- Did I say what I was thinking?
- Did I express how I feel?

WORD CHOICE
- Will my reader understand my words?
- Did I use words I love?
- Are my words interesting?
- Can I picture it?

SENTENCE FLUENCY
- Is my paper easy to read out loud?
- Do my sentences begin in different ways?
- Are some sentences long and some short?

CONVENTIONS
- Did I use paragraphs?
- Is it easy to read my spelling?
- Did I use capital letters in the right place?
- Are periods, commas, exclamation marks, and quotation marks in the right places?

How Do Ecosystems Change Naturally?

Read pages B92 to B95 in your textbook. Then read each question that follows. Decide which is the best answer to each question. Mark the letter for that answer.

HINT Look for words in italics that introduce the three stages of succession.

1. Which one is an example of primary succession?

    A Flowering plants take root in the soil.

    B Lichens grow near a glacier.

    C Spruce and hemlock trees grow.

    D A prairie develops.

HINT Use surrounding words to help you figure out the meaning of a word.

2. Read this sentence from the lesson.

    Pioneer plants are the first plants to invade a bare area.

    What does the word *invade* mean?

    A decorate

    B leave

    C damage

    D enter

HINT What kinds of ecosystems show examples of primary succession?

3. Why is Glacier Bay, Alaska, a perfect place for the study of primary succession?

    A Its glaciers keep melting, leaving new areas of exposed rock.

    B Its cold climate is good for growing plants.

    C The area has many deciduous forests.

    D The area has a huge variety of plant and animal species.

HINT Can you locate a word in the lesson that means "changing from one form to another"?

4. What do scientists call a community of plants that develops between a pioneer and a climax community?

    A the mossy stage

    B secondary succession

    C a forest community

    D a transitional community

HINT Reread the lesson to find facts and examples concerning secondary succession.

5. When would secondary succession take place?

    A after a glacier melts

    B after a forest fire

    C after a volcanic island emerges from the sea

    D when organic matter becomes trapped in dense moss

HINT Remember to determine the meaning of a word from its context—how it is used in a sentence.

6. Read this sentence from the lesson.

    Eventually, most ecosystems reach a stable stage, called the climax community.

    Which word has a meaning similar to *stable?*

    A temporary   **C** lasting

    B fast   **D** healthy

Name _____

Date _____

Reading
Comprehension

How Do People Change Ecosystems?

Read pages B98 to B101 in your textbook. Then read each question that follows. Decide which is the best answer to each question. Mark the letter for that answer.

HINT Look for important words in bold type.

7. What is acid rain?

   A the sulfuric acid that is given off from smokestacks

   B the acids that form when nitrogen oxides and sulfur dioxide mix with water vapor and then fall to Earth as rain

   C combinations of pesticides that are sprayed on crops from planes and then carried by the wind

   D rain that is produced when the upper atmosphere becomes heated by pollutants rising from smokestacks and car exhausts

HINT Use both the photograph and the text on page B100 to help you answer this question.

8. Strip mining is so damaging because —

   A valuable minerals that cannot be replaced are being removed from the Earth

   B huge holes are left in the landscape which weaken the underlying bedrock

   C vast areas of topsoil and overlying rock are completely stripped away, destroying ecosystems

   D during the mining process, rivers and streams are polluted and the area's natural resources are used up

HINT Summarize each section to help you determine the lesson's main idea.

9. What is the **MOST** important idea of this lesson?

   A Farming, strip mining, industry, and other human activities are slowly damaging fragile ecosystems.

   B Ecosystems can be saved from destruction if we take several corrective steps.

   C The governments of Canada and the United States have passed several measures to control air and water pollution.

   D Human activities are damaging thousands of acres of forest land.

HINT Use context clues to help you determine the meaining of words.

10. Pollution is —

   A anything that smells bad

   B animal wastes

   C any waste product that damages an ecosystem

   D any waste product produced by factories

HINT Study the map key to help you answer this question.

11. Study the map on page B99. The area that most suffers from acid rain is —

   A the Pacific coast

   B the south Atlantic coast

   C Western Canada

   D the Eastern United States

Name _____

Date _____

How Can People Treat Ecosystems More Wisely?

Read pages B104 to B107 in your textbook. Then read each question that follows. Decide which is the best answer to each question. Mark the letter for that answer.

HINT Prefixes can change the meaning of a word.

12. The prefix *re-* means —

   A turn   **C** against

   B fix   **D** again

HINT Read captions of scientific information carefully.

13. How is a landfill different from a dump?

   A A landfill has only soil. A dump has garbage.

   B In a landfill, each layer of garbage is covered with a layer of clay soil.

   C In an open dump there is no drainage, and the dump floods. A landfill is designed so that runoff from the garbage drains into the soil.

   D A landfill is a dump in which chemicals are added to break down the garbage.

HINT Important words sometimes appear in italic type.

14. What is a catalytic converter in a car?

   A a device that turns some of the poisonous exhaust gases into water and carbon dioxide

   B an air purifier for the harmful exhaust gases entering your car

   C a device that allows your car's air conditioner to use a harmless coolant instead of the pollutant, freon

   D a substance that is added to gasoline so that it burns cleaner

HINT You will need to apply information from this lesson to answer this question.

15. After the picnic, Arlen had three dirty napkins, six chicken bones of various sizes, an empty juice can, an apple core, the stem from a small bunch of grapes, and the plastic container from his salad. Which should he recycle?

   A napkins, juice can, plastic container

   B juice can

   C juice can, apple core, grape stem, plastic container, napkins

   D juice can, plastic container

HINT When you conserve, you use less of something.

16. These are the dials on an air conditioner. Which dials are set to cool the room while conserving the most electricity?

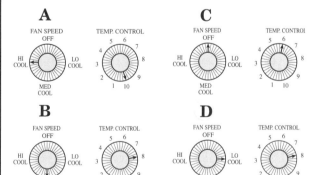

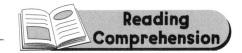
How Can People Help Restore Damaged Ecosystems?

Read pages B110 to B113 in your textbook. Then read each question that follows. Decide which is the best answer to each question. Mark the letter for that answer.

HINT Prefixes change the meaning of a word.

17. Read these sentences from the lesson.

    For example, in many freshwater ecosystems, the bodies of fish contain high levels of poisons called PCBs. When carnivores eat these fish, PCBs enter their bodies. And when the fish die and decompose, the PCBs reenter the water.

    What does the word *decompose* mean?

    A to float

    B to break down

    C to leave

    D to join

HINT Reread the passage to look for important information.

18. What is the main reason it takes so much work to restore a damaged freshwater ecosystem?

    A It is difficult to remove poisons from the water and the animals who live in it.

    B It is always an expensive project.

    C Many workers are needed for such a job.

    D It is almost impossible to clean water that is polluted.

HINT You can check your understanding of key ideas by rereading a passage.

19. Which is an example of an ecosystem that was destroyed, then replaced?

    A the Hudson River in New York

    B the Florida Everglades

    C the wetlands of Arcata, California

    D prairie lawns

HINT Which heading introduces this information?

20. How do wetlands help to purify water?

    A The wetlands can be turned into parks.

    B The marine animals in the wetlands help purify the water.

    C Scientists treat waste water in the wetlands.

    D The plants in wetlands act as natural filters.

HINT What are things people can do on their own to help the environment?

21. According to the lesson, how can gardeners help to restore natural ecosystems?

    A Gardeners can use different kinds of pesticides.

    B Gardeners can replace lawns with the kinds of plants that originally grew in the area.

    C Gardeners can ask local governments for help.

    D Gardeners can plant traditional lawns.

Unit B, Chapter 4

Base your answers on the information in this chapter. Read all parts to each question before you begin.

People's actions change ecosystems. Sometimes, these actions damage ecosystems. Describe three ways that people damage ecosystems.

HINT What kinds of human activities can hurt the environment?

_____

_____

_____

_____

_____

_____

_____

_____

_____

Burning fossil fuels can produce acid rain. Explain how acid rain affects ecosystems.

HINT Use the map and photograph on page B99 to give you ideas for your response.

_____

_____

_____

_____

_____

_____

_____

Treating Ecosystems Wisely

Ecosystems change naturally through a process known as succession. Sometimes this change is rapid. More often, however, succession takes thousands of years. Sometimes rapid changes to an ecosystem are caused by human activities. Sometimes these activities are very harmful.

In this chapter, you learned about protecting and preserving ecosystems. Design a pamphlet that would encourage other students to use resources wisely. Include at least five things people could do. Remember to include these key words: reduce, reuse, and recycle.

Use this page for prewriting or planning activities. Then write your response on a separate sheet of paper.

Writer's Checklist

IDEAS

- Is my message clear?
- Do I know enough about my topic?
- Have I included interesting details?

ORGANIZATION

- Does my paper start out with a bang?
- Did I tell things in the best order?
- At the end does it feel finished and make you think?

VOICE

- Does this writing really sound like me?
- Did I say what I was thinking?
- Did I express how I feel?

WORD CHOICE

- Will my reader understand my words?
- Did I use words I love?
- Are my words interesting?
- Can I picture it?

SENTENCE FLUENCY

- Is my paper easy to read out loud?
- Do my sentences begin in different ways?
- Are some sentences long and some short?

CONVENTIONS

- Did I use paragraphs?
- Is it easy to read my spelling?
- Did I use capital letters in the right place?
- Are periods, commas, exclamation marks, and quotation marks in the right places?

Name _____

Date _____

 **Math Practice**

Unit B, Chapter 1

Read each question and choose the best answer. Mark the letter for that answer.

Use the circle graph below for problems 1 and 2.

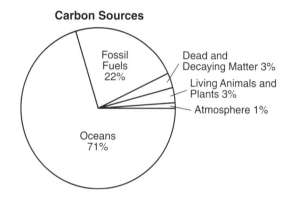

Carbon Sources

Fossil Fuels 22%
Dead and Decaying Matter 3%
Living Animals and Plants 3%
Atmosphere 1%
Oceans 71%

1. Where is the largest amount of carbon found on Earth?

   A fossil fuels

   B oceans

   C living animals and plants

   D atmosphere

2. What percent more carbon do fossil fuels contribute to Earth than living animals and plants?

   A 19%   **C** 22%

   B 21%   **D** 68%

3. Earth's atmosphere is 78% nitrogen. Determine the percent of Earth's atmosphere left for other substances.

   A 2%   **C** 22%

   B 12%   **D** 32%

4. A glass containing 100 milliliters of water sat in the sun for two days. Due to evaporation, the glass contained only 40.5 milliliters after the two days. How many milliliters of water evaporated?

   A 40.5 mL   **C** 60.5 mL

   B 59.5 mL   **D** 100 mL

Use the bar graph below for problems 5 and 6.

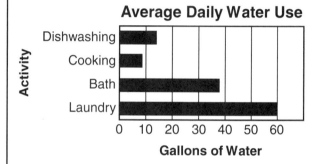

Average Daily Water Use

Activity: Dishwashing, Cooking, Bath, Laundry
Gallons of Water: 0 10 20 30 40 50 60

5. Which of the following sentences is TRUE?

   A Cooking uses more water than taking a bath.

   B A bath uses less water than dishwashing.

   C Doing laundry uses more water than any other activity.

   D Dishwashing uses less water than any other activity.

6. Determine how many more gallons of water doing the laundry requires than dishwashing.

   A 12 gal   **C** 24 gal

   B 20 gal   **D** 46 gal

Unit B, Chapter 2

Read each question and choose the best answer. Mark the letter for that answer.

7. For an experiment on a small ecosystem, the students are asked to mark off a plot of grass that is 10 feet long and $5\frac{1}{2}$ feet wide. What is the area of the ecosystem?

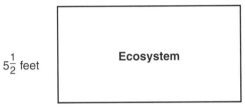

5½ feet — Ecosystem — 10 feet

A 15 sq. ft   **C** 55 sq. ft

B 30 sq. ft   **D** 100 sq. ft

Use the bar graph below for problems 8 and 9.

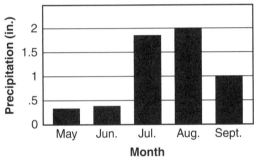

Southern Arizona / Tucson

8. Determine which month has the highest amount of rainfall.

A May   **C** July

B June   **D** August

9. Which of the following statements is reasonable?

A July has less rainfall than September.

B June has less rainfall than September.

C August has less rainfall than September.

D September has the greatest amount of rainfall.

Use the table below for problems 10 and 11.

Insects Found on Three Hunts

Hunt	Red	Blue	Green	Yellow
1	3	0	1	2
2	2	1	5	7
3	0	4	3	1

10. Choose the bar graph that best represents the data in the table.

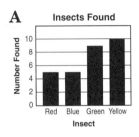 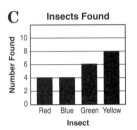

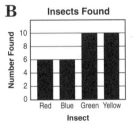 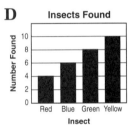

11. Which of the following statements is reasonable?

A Red insects are most commonly found in the ecosystem used during the 3 hunts.

B Blue insects are most commonly found in the ecosystem used during the 3 hunts.

C Green insects are most commonly found in the ecosystem used during the 3 hunts.

D Yellow insects are most commonly found in the ecosystem used during the 3 hunts.

Math Practice

Unit B, Chapter 3

Read each question and choose the best answer. Mark the letter for that answer.

Use the table below for problems 12 and 13.

Amount of Rainfall

Biome	Yearly Precipitation (cm)
Tropical rain forest	250 cm
Desert	10 cm
Tundra	25 cm

12. Choose the bar graph that best represents the data in the table.

A

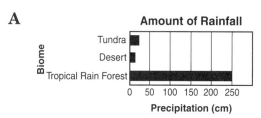

B

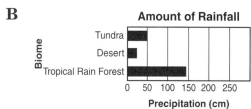

C

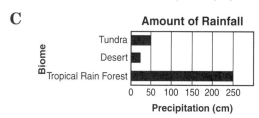

D

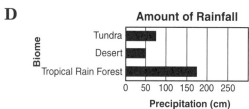

13. According to the table, would you expect to find more trees in the tundra or in a tropical rain forest?

   A tundra

   B tropical rain forest

   C They would have the same amount of trees.

   D cannot be determined

Use the circle graph below for problems 14 and 15.

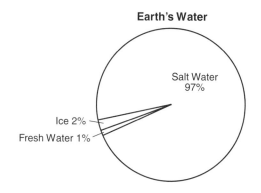

14. If you were to close your eyes and point to a place on a globe, would you be more likely to point to a body of salt water or fresh water?

   A fresh water

   B salt water

   C ice

   D They would be equally likely.

15. What percent more salt water makes up Earth's water than ice?

   A 2%   **C** 96%

   B 95%   **D** 97%

16. What would be the best unit of measure to find the amount of water in a glass?

   A tons   **C** centimeters

   B inches   **D** milliliters

17. Desert bushes such as the mesquite and creosote grow roots up to 15 meters long to reach underground water. How many centimeters long are the roots?

   A 15 cm   **C** 1500 cm

   B 150 cm   **D** 15,000 cm

Unit B, Chapter 4

Read each question and choose the best answer. Mark the letter for that answer.

18. What is the volume of the box below?

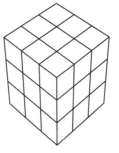

- **A** 3 cubic units
- **B** 9 cubic units
- **C** 12 cubic units
- **D** 27 cubic units

19. Choose the bar graph that best represents the data in the table below.

Acidity Levels

Item	Acidity Scale level
Lemon juice	2.3
Vinegar	3.3
Sea water	8.0

A
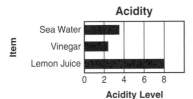

B
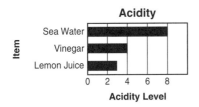

C
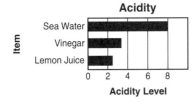

D
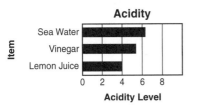

20. If you sprinkle a plant with 50 milliliters of water every day for 100 days, how many liters of water will you use?

- **A** 5 L
- **B** 50 L
- **C** 500 L
- **D** 5000 L

21. There are 20 layers of soil, each 15.7 centimeters thick, in a landfill. How many centimeters of soil are there all together?

- **A** 35 cm
- **B** 314 cm
- **C** 350 cm
- **D** 3000 cm

22. The Florida Everglades once covered about 4 million acres. Today it only covers half as much. How many acres is that?

- **A** 2 million acres
- **B** 4 million acres
- **C** 6 million acres
- **D** 8 million acres

23. Which unit would be best to use to measure the area of a wetland?

- **A** square centimeters
- **B** square inches
- **C** square feet
- **D** square kilometers

Math Practice

Unit B, Review

Read each question and choose the best answer. Mark the letter for that answer.

24. A glass containing 1 liter of water sat in the sun for 5 days. Due to evaporation, the glass contained only 150 milliliters at the end of the 5 days. How many milliliters evaporated?

A 149 mL　　**C** 850 mL

B 500 mL　　**D** 1000 mL

25. For an experiment on a small ecosystem, the students are asked to mark off a plot of grass that is 3 meters long and 5 meters wide. What is the area of this ecosystem?

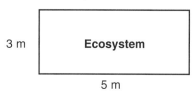

3 m | Ecosystem

5 m

A 2 square meters

B 8 square meters

C 15 square meters

D 19 square meters

26. During an experiment 3 red insects, 2 blue insects, and 1 yellow insect were found in an ecosystem. What is the probability that the first insect found was a red one?

A $\frac{1}{5}$　　**C** $\frac{1}{3}$

B $\frac{1}{4}$　　**D** $\frac{1}{2}$

27. What is the volume of the box of soil below?

A 12 cubic units

B 36 cubic units

C 48 cubic units

D 55 cubic units

Use the table below for problems 28 and 29.

Amount of Rainfall

Biome	Yearly Precipitation (cm)
Grasslands	27 cm
Deciduous forest	150 cm
Taiga	45 cm

28. Choose the bar graph that best represents the data in the table.

A
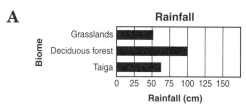

B
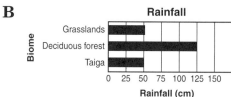

C
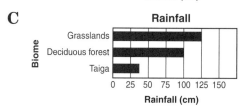

D
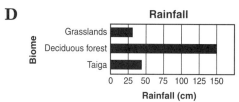

29. According to the table, which of the following statements is reasonable?

A Deciduous forests have more yearly precipitation than grasslands.

B Taigas have more yearly precipitation than deciduous forests.

C Grasslands have more yearly precipitation than taigas.

D Grasslands have more yearly precipitation than all other biomes.

What Processes Change Landforms?

Read pages C6 to C11 in your textbook. Then read each question that follows. Decide which is the best answer to each question. Mark the letter for that answer.

HINT Context clues can help you determine the meaning of unfamiliar words.

1. What are flood plains?

    A large, sandy areas where river water has receded

    B small ponds that are formed when rivers overflow their banks

    C flat areas along riverbanks where sediment is deposited

    D old, dried out lake beds

HINT Skim the text under the heading "Ice" to answer this question.

2. Which is **NOT** a characteristic of glaciers?

    A They are permanent landforms.

    B They move.

    C They are made almost entirely of ice.

    D They are agents of erosion.

HINT Use the illustration on page C9 to help you answer this question.

3. A sinkhole forms when —

    A limestone is weakened by the carbonic acid in rainwater that seeps into the ground

    B the weight of mud and sand cause the ground to give way

    C pollutants in underground streams erode the bedrock and cause part of the ground to collapse

    D flowing water gradually eats away at the ground, causing a depression

HINT Important information often is found at the beginning of a section.

4. What is the main idea of the section headed "New Landforms"?

    A Glaciers are major forces for forming new landforms.

    B New islands can be formed by volcanic eruptions.

    C Long Island and the Hawaiian islands are formed by different processes.

    D Erosion and deposition can change landforms or produce new ones.

HINT Look for details when identifying facts from a lesson.

5. Which of the following is **NOT** a FACT?

    A Glaciers erode sediment.

    B Glaciers do not move.

    C When glaciers meet, moraines are formed.

    D Glaciers can change landforms.

HINT Think about the agents of weathering and erosion that are most active in dry environments.

6. The sand dunes on a beach cannot form without —

    A the activities of sand crabs and other marine life

    B the deposition of shells

    C the action of wind

    D wave action

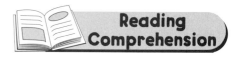
What Causes Mountains, Volcanoes, and Earthquakes?

Read pages C14 to C19 in your textbook. Then read each question that follows. Decide which is the best answer to each question. Mark the letter for that answer.

HINT Illustrations can help clarify text information.

7. Which is the correct path to Earth's center?

   A mantle, outer core, inner core, crust

   B crust, mantle, outer core, inner core

   C outer core, crust, mantle, inner core

   D outer core, inner core, mantle, crust

HINT Use the map on page C17 to help you answer this question.

8. Which house is furthest from the ring of fire?

   A Dave's Dwelling: on the west coast of North America

   B Hattie's Homestead: in the western Pacific, just off the coast of Asia

   C Pete's Pad: on the southwest coast of Australia

   D Lilly's Lodge: on the northwest coast of South America

HINT Read carefully to locate specific details about a subject.

9. Which of the following is an extinct volcano?

   A the Himalayas

   B Kilauea

   C the Grand Tetons

   D Kure Atoll

HINT Headings can help you locate specific information in the text.

10. The movement of the African and Arabian plates is causing —

   A mountains to form in eastern Africa

   B volcanoes to erupt in eastern Africa

   C an ever-widening crack that will someday split Africa

   D submarine ridges to form off the African coast

HINT Skim the lesson to review important points.

11. Which of the following is **NOT** TRUE?

   A The Mid-Atlantic Ridge is a mountain chain located in central Europe.

   B Lava is magma that reaches the surface.

   C Mountains are formed when plates push against each other.

   D The Hawaiian Islands are actually the tops of volcanoes.

HINT Compare these diagrams to the one on page C18.

12. Which shows how seismic waves move?

   A

   B

   C

   D

How Has Earth's Surface Changed?

Read pages C22 to C25 in your textbook. Then read each question that follows. Decide which is the best answer to each question. Mark the letter for that answer.

HINT Key information is often found at the beginning of a lesson or passage.

13. What was Laurasia?

A a legendary island like Atlantis

B a continent that was completely swallowed up by the sea

C one of two large continents that broke apart and formed the present continents

D the name given to a large land mass thought to exist in the Pacific Ocean during the time of the dinosaurs

HINT What can scientists tell from the position of rock layers?

14. Which fossil is the oldest?

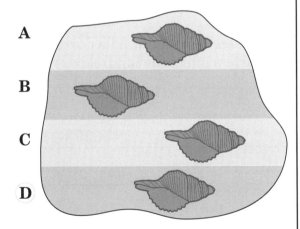

HINT Important information can sometimes be found in captions.

15. A person who studies fossils is called —

A an archeologist **C** a zoologist

B a paleontologist **D** a geologist

HINT A hypothesis is an explanation that must be tested to be proven true.

16. Which of the following is a hypothesis?

A The walls of the Grand Canyon are composed of sedimentary rock.

B The Atlantic Ocean is getting wider.

C Some fossils are the actual remains of once-living organisms.

D All the continents were once part of a large land mass.

HINT Skim the lesson to review important points.

17. What can you conclude from reading this lesson?

A If scientists can uncover the very deepest fossils, they will be able to find out what the earliest plants and animals looked like.

B Sometimes the best that scientists can do is to make educated guesses about Earth's past.

C The older the rock layers, the more fossils there are.

D The greatest geological changes have already occurred, and Earth is not likely to change much in the next several million years.

Name _____

Date _____

Unit C, Chapter 1

Base your answers on the information in this chapter. Read all parts to each question before you begin.

You know that Earth's surface has changed. Water is an important agent of these changes. Explain how water can change landforms. Include examples of the changes you describe.

HINT Use the key words in this chapter that are highlighted in yellow to help you answer the question.

_____

_____

_____

_____

_____

_____

Glaciers are thick sheets of ice that form in areas where more snow falls during the winter than melts during the summer. Tell how glaciers affect the landscape.

HINT What are the two types of glaciers?

_____

_____

_____

_____

_____

_____

_____

Demonstrating Mountain Formation

Earth's surface is constantly changing. The plates that make up Earth's crust are constantly moving. Mountains, Earth's highest landforms, are formed as the crust folds, cracks, and bends upward because of movements of Earth's plates.

Imagine that you are going to use some simple materials to demonstrate to a group of younger students how mountains form. Write step-by-step instructions detailing how you would perform these demonstrations and the type of mountain formation each demonstration illustrates.

Use this page for prewriting or planning activities. Then write your response on a separate sheet of paper.

Writer's Checklist

IDEAS

- Is my message clear?
- Do I know enough about my topic?
- Have I included interesting details?

ORGANIZATION

- Does my paper start out with a bang?
- Did I tell things in the best order?
- At the end does it feel finished and make you think?

VOICE

- Does this writing really sound like me?
- Did I say what I was thinking?
- Did I express how I feel?

WORD CHOICE

- Will my reader understand my words?
- Did I use words I love?
- Are my words interesting?
- Can I picture it?

SENTENCE FLUENCY

- Is my paper easy to read out loud?
- Do my sentences begin in different ways?
- Are some sentences long and some short?

CONVENTIONS

- Did I use paragraphs?
- Is it easy to read my spelling?
- Did I use capital letters in the right place?
- Are periods, commas, exclamation marks, and quotation marks in the right places?

What Are Minerals?

Read pages C36 to C39 in your textbook. Then read each question that follows. Decide which is the best answer to each question. Mark the letter for that answer.

HINT What are the characteristics of minerals?

1. Which of the following statements is **NOT** a **FACT**?

   A All minerals are solids.

   B All minerals are made from material that was never alive.

   C Minerals can be made in a laboratory.

   D Diamonds are minerals.

HINT Look for important information in tables and charts.

2. You found an interesting rock while you were hiking. It's very shiny, and it left a grey streak when you tested it. Which of the following is it most likely to be?

   A magnetite

   B hematite

   C galena

   D chalcopyrite

HINT Hardness is a mineral's ability to resist being scratched.

3. A penny is —

   A harder than quartz

   B softer than gypsum

   C as hard as calcite

   D as hard as a fingernail

HINT Important vocabulary is indicated with yellow highlighting.

4. The word *luster* means —

   A dullness

   B hardness

   C shine

   D streak

HINT Skim the lesson for information about diamonds.

5. You looked up the word *diamond* in the index of a book about minerals, but it wasn't there. What other topic should you check before you try another book?

   A mantle

   B crystals

   C hardness

   D quartz

HINT Use context clues to determine the meaning of "refined."

6. What happens when a mineral is refined?

   A It is made pure by removing other substances from it.

   B Other substances are added to it.

   C It is changed into another form.

   D It is turned into something useful, such as wallboard or copper wire.

Name _____

Date _____

What Are Rocks?

Read pages C42 to C47 in your textbook. Then read each question that follows. Decide which is the best answer to each question. Mark the letter for that answer.

HINT Under which heading will you find this information?

7. Which is an igneous rock?

   A sandstone

   B granite

   C shale

   D gneiss

HINT Important words sometimes are indicated with italic type.

8. Why does pumice have little holes in it?

   A The holes are made by small organisms.

   B The rock is soft, and it wears away very easily.

   C Pumice is actually a sea sponge that has become a rock.

   D The holes are caused by gases that escaped as the rock cooled.

HINT Arrows often indicate movement.

9. In the diagram on page C45, what do the arrows represent?

   A the movement of rock

   B wind flow and direction

   C ocean currents

   D the rise and fall of tides

HINT Review the lesson and identify the main idea of each section.

10. This lesson tells about all of the following EXCEPT —

    A how volcanoes form

    B why some roofs are made out of slate

    C how crossbeds are formed

    D how sandstone got its name

HINT What does the word "conglomerate" mean?

11. When a company is referred to as "a big conglomerate," what does that mean?

    A The company is made up of many smaller companies.

    B The company is old and established.

    C The company has many employees.

    D The company has branch offices in several cities.

HINT What happens to both butterflies and rocks when they undergo metamorphosis?

12. A metamorphosis is —

    A a slow build up

    B a slope

    C a change

    D an explosion

Name _____

Date _____

What Is the Rock Cycle?

Read pages C50 to C53 in your textbook. Then read each question that follows. Decide which is the best answer to each question. Mark the letter for that answer.

HINT Follow the arrows to read the captions in the correct order.

13. What does the diagram on pages C50 and C51 show?

    A what happens when a volcano erupts

    B how igneous rock becomes sedimentary rock and then finally metamorphic rock

    C how rock is eroded and carried long distances

    D how the rocks produced when a volcano erupts help to build up the level of the sea floor

HINT Trace the path of one color of arrows at a time and determine what each color represents.

14. The diagram on page C52 shows you that quartzite —

    A andesite, and sandstone are all the same

    B can become part of sandstone or andesite, depending on the conditions

    C breaks up to produce sandstone and andesite

    D is always the first type of rock formed during the rock cycle

HINT Use context clues to help you determine the meaning of unknown words.

15. What does it mean for a rock to be compacted?

    A enlarged   **C** cut in half

    B compressed   **D** assembled

HINT Read for details in the captions.

16. Study the diagram on pages C52 and C53. Which of the following is **NOT** a FACT?

    A Quartzite can become an igneous rock.

    B Sandstone can melt completely and form quartzite.

    C Andesite can become a metamorphic rock.

    D Andesite can be broken down and compacted to form sedimentary rock.

HINT Make a list of the processes as you study the diagram on pages C52 and C53.

17. Some of the processes in the rock cycle are —

    A erosion

    B compaction

    C cementation

    D all of these

HINT Follow the red arrow from quartzite to answer this question.

18. What must happen to change a metamorphic to an igneous rock?

    A It has to get hot enough to melt and form magma.

    B It has to sink to the bottom of the river.

    C It has to freeze and thaw.

    D Its jagged edges need to be rounded off.

Unit C, Chapter 2

Base your answers on the information in this chapter. Read all parts to each question before you begin.

In order to be a mineral, a material must be a solid with particles arranged in a repeating pattern called a crystal. Most minerals are made from material that was never alive. Describe at least two ways that minerals form.

HINT Under which heading are you likely to find this information?

_____

_____

_____

_____

_____

Scientists use several properties to identify minerals. Some of these properties are listed here. Describe three properties of minerals. Give an example of each property in your response.

HINT Use the text and illustrations on pages C36 and C37 to find details and examples to include in your answer.

_____

_____

_____

_____

_____

_____

_____

_____

Name _____

Date _____

Rock Formation

Earth is made mostly of rocks. Rocks are constantly changing through a process called the rock cycle. Normally the change takes place slowly over thousands of years as new rocks form from old rocks through natural processes. Some rocks form deep inside Earth. Others form on Earth's surface.

Write an article for a science magazine that compares and contrasts the formation of igneous and sedimentary rocks.

Use this page for prewriting or planning activities. Then write your response on a separate sheet of paper.

Writer's Checklist

IDEAS

- Is my message clear?
- Do I know enough about my topic?
- Have I included interesting details?

ORGANIZATION

- Does my paper start out with a bang?
- Did I tell things in the best order?
- At the end does it feel finished and make you think?

VOICE

- Does this writing really sound like me?
- Did I say what I was thinking?
- Did I express how I feel?

WORD CHOICE

- Will my reader understand my words?
- Did I use words I love?
- Are my words interesting?
- Can I picture it?

SENTENCE FLUENCY

- Is my paper easy to read out loud?
- Do my sentences begin in different ways?
- Are some sentences long and some short?

CONVENTIONS

- Did I use paragraphs?
- Is it easy to read my spelling?
- Did I use capital letters in the right place?
- Are periods, commas, exclamation marks, and quotation marks in the right places?

Name _____

Date _____

How Can You Observe and Measure Weather Conditions?

Read pages C64 to C69 in your textbook. Then read each question that follows. Decide which is the best answer to each question. Mark the letter for that answer.

HINT Read captions carefully for important details.

1. If you wanted to know how humid it was outside, you would check a —

   A hygrometer

   B barometer

   C thermometer

   D rain gauge

HINT Important information often is found at the beginning of a lesson.

2. Captain Grover just returned from the planet X-25. When he was asked what it was like on X-25, the captain said, "Neat! The sky was pitch black and there was no weather." What can you conclude about the planet?

   A It did not circle a star.

   B It was much smaller than Earth.

   C It did not have a rocky surface.

   D It had no atmosphere.

HINT Reread the lesson looking for important details.

3. Which of the following is **NOT** TRUE?

   A As air gets warmer, air pressure increases.

   B Cold air has less humidity than warm air.

   C The lowest layer of the atmosphere has more water than the uppermost layer.

   D Earth's surface cools faster on clear nights than on cloudy ones.

HINT Reread page C67 to help you answer this question.

4. The water cycle describes how —

   A water moves from lakes and streams to the oceans and then back again

   B salt is removed from ocean water through evaporation

   C water evaporates from Earth, condenses, and returns as precipitation

   D currents in the upper atmosphere affect the advance of weather fronts

HINT Use the diagram on page C68 to help you answer this question.

5. Who went through a stratus cloud?

   A Captain McDuffy, flying his plane at 10 km

   B Captain Buzz, sailing his hot-air balloon at an altitude of 5 km

   C Felicia, who did a loop-the-loop at 2.5 km in her stunt plane

   D Raul, who skated through a patch of ground fog

HINT Use section headings to help you locate important information.

6. A high pressure area is usually —

   A a mass of cold air

   B a mass of warm air

   C a bank of cumulus clouds

   D the calm area around a storm cloud

What Causes Weather?

Read pages C72 to C77 in your textbook. Then read each question that follows. Decide which is the best answer to each question. Mark the letter for that answer.

HINT Compare the diagram and the text to help you determine the section's main idea.

7. What is the main idea of the section headed "Uneven Heating"?

   A Earth's surface absorbs about 50% of the sun's energy.

   B The atmosphere absorbs some of the sun's energy and reflects some of it back into space.

   C Earth absorbs and reflects the sun's energy unevenly.

   D Winds are caused by the heating and cooling of Earth.

HINT Arrows in diagrams often indicate direction of movement.

8. The global winds blowing over the southern tip of South America are —

   A prevailing westerlies

   B northeast trades

   C southeast trades

   D Polar easterlies

HINT Some terms are defined in captions.

9. What is a land breeze?

   A wind that flows across the land

   B cooler air from the land flowing toward the sea

   C cooler air from the sea moving toward the land

   D warm air from the land rising at night

HINT Use the key to help you interpret this map.

10. Study the weather map on page C76. On March 23rd, the weather in Seattle was —

    A cold and snowy

    B warm and rainy

    C mostly sunny

    D cool and rainy

HINT An inference is a conclusion based on observed facts.

11. By studying the diagram of global winds on page C74, you can infer that —

    A winds are named for the direction from which they blow

    B all winds blow toward the equator

    C all prevailing winds blow at the same speed

    D winds flow around Australia in a counterclockwise motion

HINT Use context clues to help you determine the meaning of unfamiliar terms.

12. An updraft is —

    A the horizontal movement of air

    B a strong, sudden wind

    C the upward movement of warm air

    D any spiral-shaped wind

Name _____

Date _____

What Is Climate and How Does It Change?

Read pages C80 to C85 in your textbook. Then read each question that follows. Decide which is the best answer to each question. Mark the letter for that answer.

HINT Use the color coding on the map on page C82 to help you answer this question.

13. Which of the following areas have a similar climate?

   A the northern half of South America and the southern tip of Africa

   B the eastern half of North America and the northern part of Eurasia

   C Australia and the islands in the Pacific Ocean

   D the western United States and central Africa

HINT The main idea is often found at the beginning of a section.

14. What is the main idea of the section headed "World Climates"?

   A Many factors affect an area's climate, so no two places have exactly the same climate.

   B There are five climate zones.

   C Each climate zone has its own particular plants and animals that are adapted to that climate.

   D Ocean currents greatly affect climate.

HINT Look for highlighted vocabulary words in each lesson.

15. What is El Niño?

   A a weather change that signals the start of an ice age

   B a change in the Pacific Ocean currents

   C a short-term climate change due to a change in ocean currents

   D a powerful wind that blows onshore from the Pacific Ocean and affects weather

HINT Use the key to help you interpret this map.

16. What can you conclude by studying the diagram on page C83?

   A Earth's temperature does not remain constant but instead rises and falls in great cycles.

   B Earth's temperature is about as warm today as it was 350,000 years ago.

   C So far, Earth's recent rise in temperature seems normal and not a cause for alarm.

   D Earth's temperature will never rise above the level it was 125,000 years ago.

HINT Reread the section "Humans Affect Climate" to find the answer to this question.

17. According to the author, what will happen if Earth's average temperature rises just a few more degrees?

   A We will enter another ice age.

   B The amount of carbon dioxide in the atmosphere will increase.

   C The polar ice caps will begin to melt, and coastal cities will flood.

   D Massive evaporation will occur, exposing vast sections of the seafloor.

Unit C, Chapter 3

Base your answers on the information in this chapter. Read all parts to each question before you begin.

You experience weather every day. Weather occurs in the layer of Earth's atmosphere that is closest to the surface. Explain why Earth's weather occurs only in the troposphere.

HINT Which heading introduces this information?

_____

_____

_____

_____

_____

_____

You know that weather changes from day to day. But what causes these changes? Explain the main factor that causes weather to change. Include in your explanation how you can predict a change in the weather.

HINT Look under the heading "Air Pressure" to find this information.

_____

_____

_____

_____

_____

_____

_____

_____

Weather Letters

The weather in any area may change from day-to-day. However, the same pattern of weather usually repeats from year-to-year. This repeating weather pattern is part of an area's climate. People often choose to vacation in an area with a climate that is different from the one at home. For example, many people who live where winters are cold choose to vacation in areas where winters are warm.

Imagine you are on vacation at a particular place. Write a letter to a friend describing the weather there. Include the temperature, the wind speed, and the wind direction. Describe any local winds that occur in that area. Also include information about changes you expect in the weather, and why you expect them.

Use this page for prewriting or planning activities. Then write your response on a separate sheet of paper.

Writer's Checklist	
IDEAS • Is my message clear? • Do I know enough about my topic? • Have I included interesting details?	**WORD CHOICE** • Will my reader understand my words? • Did I use words I love? • Are my words interesting? • Can I picture it?
ORGANIZATION • Does my paper start out with a bang? • Did I tell things in the best order? • At the end does it feel finished and make you think?	**SENTENCE FLUENCY** • Is my paper easy to read out loud? • Do my sentences begin in different ways? • Are some sentences long and some short?
VOICE • Does this writing really sound like me? • Did I say what I was thinking? • Did I express how I feel?	**CONVENTIONS** • Did I use paragraphs? • Is it easy to read my spelling? • Did I use capital letters in the right place? • Are periods, commas, exclamation marks, and quotation marks in the right places?

What Are the Oceans Like?

Read pages C96 to C99 in your textbook. Then read each question that follows. Decide which is the best answer to each question. Mark the letter for that answer.

HINT Use the table on page C96 to help you answer this question.

1. Which is **NOT** a true statement about Earth's oceans?

   A The Arctic is the smallest ocean.

   B The Atlantic is larger than the Indian Ocean.

   C The Pacific Ocean has more surface area than all the other oceans combined.

   D The Atlantic is about half the size of the Pacific.

HINT Use context clues to help you determine the meaning of words.

2. What does the term *salinity* mean?

   A saltiness

   B density

   C low temperature

   D area of high evaporation

HINT Reread page C98 to locate details about the ocean floor.

3. The deepest part of the ocean is the —

   A continental shelf

   B continental slope

   C the Mariana Trench

   D the Mid-Atlantic Ridge

HINT Both the diagram and the text on pages C98 and C99 will help you answer this question.

4. On which part of the ocean floor would you find ridges, volcanoes, and trenches?

   A the continental shelf

   B the abyssal plain

   C the continental slope

   D on all of these features

HINT Captions as well as the main text can provide important information.

5. Why did the *Trieste's* cabin need to be built so strongly?

   A to withstand the tremendous pressure of deep water

   B to keep out deep-sea creatures

   C to provide protection from salt

   D to allow oceanographers to cross the Pacific ocean

HINT Reread the text looking for details to help you answer this question.

6. You fall asleep on an ocean-going vessel. You wake up in an ocean that is mostly covered with ice. You are probably in the —

   A Indian Ocean

   B Arctic Ocean

   C Atlantic Ocean

   D Pacific Ocean

How Do Ocean Waters Move?

Read pages C102 to C107 in your textbook. Then read each question that follows. Decide which is the best answer to each question. Mark the letter for that answer.

HINT You can find this information under the heading "Other Kinds of Waves."

7. Captain Kluug was fourteen miles off-shore when he radioed that a massive tsunami had just lifted his boat thirty feet into the air. How did the Coast Guard know he was lying?

  A Tsunamis never grow to thirty feet.

  B Tsunamis do not form in the open ocean. They only form near the shore.

  C In the open ocean, tsunamis are usually too small to be felt.

  D The tsunami wouldn't have lifted Kluug's boat; it would have sucked it down like a whirlpool.

HINT Keys help you interpret maps.

8. A current that brings cold water down from the north is the —

  A North Atlantic Drift

  B Peru current

  C West Australia current

  D Canary current

HINT Times in this table are given using a 24-hour clock.

9. According to the tide table on page C106, the highest tide occurred on —

  A January 19 at 12:34 in the morning

  B January 19 at 12:47 in the afternoon

  C January 21 at 1:58 in the morning

  D January 17 at 11:54 at night

HINT The diagram on page C106 will help you answer this question.

10. The highest tides occur when the —

  A moon and sun are at right angles to Earth

  B moon, sun, and Earth are in a straight line

  C moon is in its first quarter phase

  D sun is between Earth and the moon

HINT Compare this diagram to the photograph on page C104.

11. What does this diagram show?

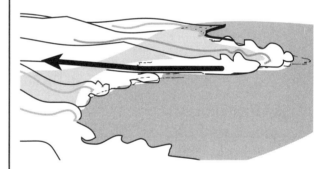

  A rip current

  B tsunami

  C storm surge

  D cold-water current

How Do Oceans Interact with the Land?

Read pages C110 to C113 in your textbook. Then read each question that follows. Decide which is the best answer to each question. Mark the letter for that answer.

HINT What is the main idea of the first section of this lesson?

12. All of the following play a major role in changing the shape of a shoreline except —

    A waves   **C** marine animals

    B weather   **D** currents

HINT Captions often contain important information.

13. What can you conclude from looking at this tide pool?

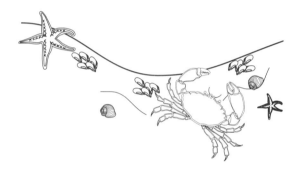

    A It is low tide.

    B It is high tide.

    C The tide pool formed very recently.

    D The animals in the tide pool have no predators.

HINT Context clues can help you determine the meaning of unfamiliar terms.

14. A longshore current moves —

    A toward the beach

    B away from the beach out to sea

    C along the shore, parallel to the beach

    D back and forth along the shore

HINT Important vocabulary often is highlighted in yellow.

15. What is a jetty?

    A a long, flat ridge of sand that forms just offshore

    B a rocky wall that sticks out into the ocean from the shore

    C a rocky ledge just under the surface of the water

    D a platform from which boats can be launched

HINT What would you expect to find in a graveyard?

16. How did the Graveyard of the Atlantic get its name?

    A It was the site of many sea caves where pirates hid stolen treasure.

    B It is a section of the Atlantic where many pirates forced their prisoners to walk the plank.

    C It was a place where many ships sank.

    D It is the furthest point from land.

HINT Reread the text and captions on page C110 to help you answer this question.

17. Which definition describes the way the word *mouth* is used in this lesson?

    A the natural opening through which food passes into the body

    B something that resembles a mouth, offering an entrance and an exit

    C the surface opening of an underground cavity

    D the place where a stream enters a larger body of water

How Do People Explore the Oceans and Use Ocean Resources?

Read pages C116 to C121 in your textbook. Then read each question that follows. Decide which is the best answer to each question. Mark the letter for that answer.

HINT Use the timeline on pages C116 and C117 to answer the question.

18. What important device for exploring the ocean was invented in 1942?

  A the submersible Alvin

  B the diving bell

  C the Aqua-lung

  D the diving suit

HINT Use context clues from the sentence to tell the meaning of the word.

19. Read the following sentence from the lesson.

They travel in small, underwater vehicles called submersibles.

What does the word *submersible* mean?

  A able to sustain breathing under water

  B able to operate on both dry land and in water

  C able to float on the water's surface

  D able to remain under water

HINT This information can be found in the passage headed "Exploring the Ocean."

20. According to the lesson, when were the first detailed studies of the oceans begun?

  A in the 1960s

  B in 1872

  C in the 1700s

  D between 1900 and 1935

HINT You can find this information under the heading "Submersibles."

21. Compared to crewed submersibles, how could crewless submersibles be useful?

  A They could remain on the ocean floor for longer periods of time.

  B They would be less expensive to manufacture.

  C They could collect better samples of marine life.

  D They would be larger than crewed submersibles.

HINT Reread the information under the heading "Using Ocean Resources."

22. Which natural resource is actually dissolved in sea water itself?

  A copper

  B nickel

  C cobalt

  D salt

HINT What is the meaning of the prefix *de-*?

23. What happens during the process of desalination?

  A Salt is dissolved in the ocean.

  B Salt is added to many foods we eat.

  C Salt is removed from sea water.

  D Salt is added to fresh water.

Unit C, Chapter 4

Base your answers on the information in this chapter. Read all parts to each question before you begin.

Most ocean waves are caused by wind. But some giant waves are caused by other forces. These waves include tsunamis, storm surges, and rogue waves. Explain the causes and effects of these three types of giant waves.

HINT Read the passage under "Other Kinds of Waves" on page C103 to find this information.

_____

_____

_____

_____

_____

_____

_____

An ocean current is a stream of water that flows like a river through the ocean. Compare surface currents, shoreline currents, and rip currents.

HINT Use the diagrams and text on pages C104 and C105 to help you write your response.

_____

_____

_____

_____

_____

_____

_____

_____

Name _____

Date _____

Ocean Currents

Surface currents move huge amounts of water through the ocean. One surface current is the Gulf Stream. This warm current flows northeast from the Caribbean Sea, past the East coast of the United States, and across the North Atlantic. Even after its long trip across the cold Atlantic, there is enough warm water in this current to warm the climate of Great Britain and northern Europe. That's why palm trees can grow along the southern coast of England.

Do some research about an ocean current such as the Gulf Stream. Find out where the current begins, where it goes, the type of water it carries, and how it affects the land it touches. Then write a short, fictional story for a classmate about the current.

Use this page for prewriting or planning activities. Then write your response on a separate sheet of paper.

Writer's Checklist	
IDEAS • Is my message clear? • Do I know enough about my topic? • Have I included interesting details?	**WORD CHOICE** • Will my reader understand my words? • Did I use words I love? • Are my words interesting? • Can I picture it?
ORGANIZATION • Does my paper start out with a bang? • Did I tell things in the best order? • At the end does it feel finished and make you think?	**SENTENCE FLUENCY** • Is my paper easy to read out loud? • Do my sentences begin in different ways? • Are some sentences long and some short?
VOICE • Does this writing really sound like me? • Did I say what I was thinking? • Did I express how I feel?	**CONVENTIONS** • Did I use paragraphs? • Is it easy to read my spelling? • Did I use capital letters in the right place? • Are periods, commas, exclamation marks, and quotation marks in the right places?

Math Practice

Unit C, Chapter 1

Read each question and choose the best answer. Mark the letter for that answer.

1. How many milliliters of liquid are in a 40-milliliter cup filled $\frac{3}{4}$ of the way?

   A 120 mL   **C** 30 mL

   B 40 mL   **D** 3 mL

2. Wind erosion can blow sand into dunes. Huge dunes as much as 100 meters high form in some deserts. How many centimeters high is this?

   A 100 cm   **C** 10,000 cm

   B 1000 cm   **D** 100,000 cm

3. As part of an experiment you are asked to put a ball of peanut butter in the freezer for a quarter of an hour. How many minutes is this?

   A 15 min   **C** 45 min

   B 30 min   **D** 60 min

4. The age of a fossil is recorded in which of the following units of measure?

   A years   **C** minutes

   B hours   **D** seconds

Use the table below for problems 5 and 6.

Major Earthquakes

Location	Magnitude
Alaska	9.1
China	8.2
California	6.8

5. How many points higher on the Richter scale was the earthquake in Alaska than the one in California?

   A 9.1   **C** 2.3

   B 6.8   **D** 1.0

6. Choose the bar graph that best represents the data in the table.

   A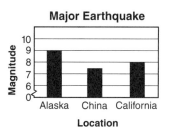

   B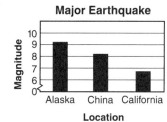

   C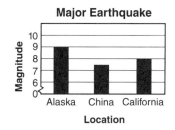

   D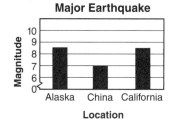

Unit C, Chapter 2

Read each question and choose the best answer. Mark the letter for that answer.

7. The pyramid of Khufu in Egypt is made of limestone, a sedimentary rock. This pyramid is the largest in the world. Each side of its base is 230 meters long. What is the area of the base?

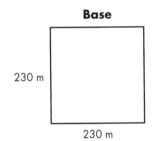

Base   **Height**

230 m

230 m

147 m

(Not drawn to scale)

  A   460 sq m

  B   920 sq m

  C   40,000 sq m

  D   52,900 sq m

8. The Khufu pyramid is 147 meters high. How many centimeters high is this?

  A   14.7 cm

  B   920 cm

  C   14,700 cm

  D   1,470 cm

9. A collecting bag contains three types of sedimentary rocks. There are 4 limestone, 2 sandstone, and 3 shale rocks in the bag. What is the probability of picking out a shale rock?

  A   2 out of 9

  B   3 out of 9

  C   4 out of 9

  D   7 out of 9

10. A lake has two rivers flowing into it. Each river deposits 1.5 centimeters of sediment in one year. How deep will the sediments be after 5 years?

  A   7.5 cm

  B   9 cm

  C   10 cm

  D   15 cm

11. What is the mass of the rock?

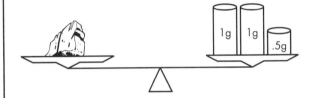

1g 1g .5g

  A   1 g

  B   1.5 g

  C   2 g

  D   2.5 g

12. If you mix together $\frac{1}{2}$ cup of laundry bluing, $\frac{2}{3}$ cup of water, and $\frac{1}{3}$ cup of ammonia for a lab, how much liquid will you have?

  A $1\frac{1}{2}$ c

  B   1 c

  C $\frac{1}{3}$ c

  D $\frac{1}{2}$ c

Math Practice

Unit C, Chapter 3

Read each question and choose the best answer. Mark the letter for that answer.

Use the table below for problems 13–15.

High Temperatures

Day	Temperature
May 1	80°F
May 2	74°F
May 3	80°F
May 4	82°F
May 5	85°F

13. Choose the line graph that best represents the data in the table.

A

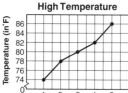

C

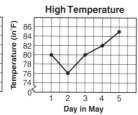

B

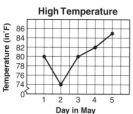

D

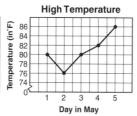

14. What is the mean temperature of the data in the table above?

A 80.2°F   **C** 74°F

B 80°F   **D** 70°F

15. What is the median temperature of the data in the table above?

A 80.2°F   **C** 74°F

B 80°F   **D** 70°F

16. Standard air pressure is 76 centimeters. How many meters is this?

A 76 m   **C** 0.76 m

B 7.6 m   **D** 0.076 m

17. Twenty-five percent of the sun's rays are absorbed and reflected by clouds. Twenty percent of the rays are absorbed and reflected by air. Five percent of the rays reflect off Earth's surface. What percent of the rays is left to be absorbed by Earth?

A 25 %   **C** 45 %

B 40 %   **D** 50 %

18. Choose the bar graph that best represents the data in the table below.

Average Daily Temperature in Winter

Place	Temperature
Coast of Maine	20°F
Outer Banks, NC	45°F
Florida	65°F

A

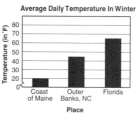

C

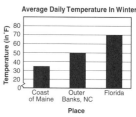

B

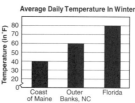

D

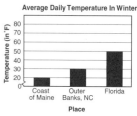

Unit C, Chapter 4

Read each question and choose the best answer. Mark the letter for that answer.

19. On a calm day, ocean waves are found to be $5\frac{1}{3}$ feet. During a storm, waves reached heights of $100\frac{1}{4}$ feet. How much higher were the waves on the stormy day than on a calm day?

A $105\frac{1}{12}$ feet   **C** $95\frac{11}{12}$ feet

B $94\frac{11}{12}$ feet   **D** $5\frac{1}{3}$ feet

20. Tsunami waves are about 62 miles long, but only 0.001 miles high. Look at the diagram below. What is the area of the rectangular region covered by this wave?

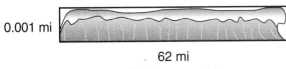

0.001 mi

62 mi

(Not drawn to scale)

A 62 square miles

B 6.2 square miles

C 0.62 square miles

D 0.062 square miles

21. On parts of Cape Hatteras ocean waves wash away 4.3 meters of beach each year. How many meters will be washed away in 10 years?

A 4.3 m

B 43 m

C 430 m

D 4300 m

Use the table below for problems 22 and 23.

Sonar Data

Location	Time (sec)
1	1.8
2	4.5
3	3.1
4	2.5

22. Choose the line graph that best represents the data in the table.

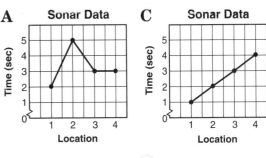

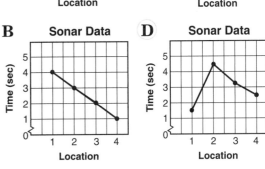

23. Find the depth of location 2 by multiplying the time by 1500 meters per second (the speed of sound in water). Then divide this product by 2.

A 6750 m   **C** 4000 m

B 4650 m   **D** 3375 m

Unit C, Review

Read each question and choose the best answer. Mark the letter for that answer.

Use the table below for problems 24–26.

Average Ocean Depths

Ocean	Depth (m)
Pacific	4,188
Indian	3,872
Atlantic	3,735
Arctic	1,038

24. Choose the bar graph that best represents the data in the table.

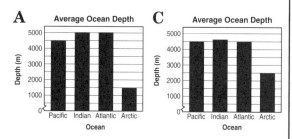

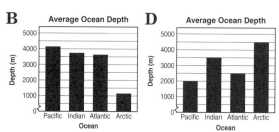

25. How much deeper is the Pacific Ocean than the Atlantic Ocean?

A 3150 m   **C** 453 m

B 2834 m   **D** 316 m

26. According to the table, which is a reasonable statement?

A The Indian Ocean is deeper than the Pacific Ocean.

B The Arctic Ocean is deeper than the Atlantic Ocean.

C The Atlantic Ocean is deeper than the Indian Ocean.

D The Pacific Ocean is the deepest.

27. Standard air pressure is 76 centimeters. A high pressure was recorded to be 80.7 centimeters. How much over standard air pressure was the high pressure?

A 4.7 cm   **C** 4 cm

B 4.3 cm   **D** 0.47 cm

28. If the trend on the line graph below continues, what would be the expected level of carbon dioxide in the year 2000?

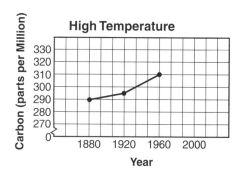

A 290 or less   **C** Around 330

B 310 or less   **D** 700

29. Once a tsunami reaches shore, it can be 62 miles long and 0.02 miles high. Look at the diagram. What is the area of the rectangular region covered by this wave?

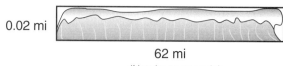

0.02 mi

62 mi
(Not drawn to scale)

A 1.24 square miles

B 12.4 square miles

C 124 square miles

D 1240 square miles

Name _____

Date _____

How Do Earth and the Moon Compare?

Read pages D6 to D11 in your textbook. Then read each question that follows. Decide which is the best answer to each question. Mark the letter for that answer.

HINT The captions on pages D6 and D7 will help you answer this question.

1. In this lesson, the word *waxing* means —

   A growing larger

   B shrinking

   C brightening

   D changing shape

HINT The word *eclipse* also means "to darken."

2. A solar eclipse occurs when —

   A there is a full moon at night

   B there is a new moon during the day

   C the moon is not visible during the day

   D the sun is below the horizon

HINT Study the diagram on page D9 to help you answer this question.

3. Rilles on the moon are similar to —

   A mountain ranges

   B volcanoes

   C flat, featureless plains

   D valleys

HINT Look for important information on page D10.

4. Which of these statements comparing Earth and the moon is **NOT** TRUE?

   A Both have craters.

   B Both are made of similar material.

   C Both have an atmosphere.

   D Neither one is made of gas.

HINT Study the photographs on pages D6 and D7.

5. Study these two phases of the moon. Which phase will you see next?

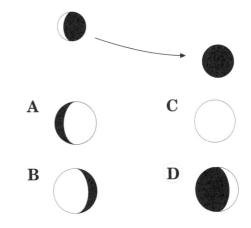

HINT Which section discusses Earth and the moon in space?

6. Which statement about the moon is **NOT** TRUE?

   A The moon rotates around Earth.

   B The moon spins on its axis.

   C The moon does not give off its own light.

   D The moon revolves around Earth.

HINT When you make an inferrence, you draw a conclusion based on what you know.

7. If the sun did not shine on the moon, the moon would —

   A not have craters

   B travel around Earth faster

   C not have phases

   D not be so close to Earth

What Else Is in the Solar System?

Read pages D14 to D19 in your textbook. Then read each question that follows. Decide which is the best answer to each question. Mark the letter for that answer.

HINT Under which heading would you find information about time?

8. Upon what is our system of telling time based?

   A the phases of the moon

   B the path Earth follows around the sun

   C Earth's 24-hour rotation

   D the date of the winter solstice

HINT Use context clues to help you determine the meaning of words.

9. Read this sentence from the lesson.

   Of the five outer planets, four are huge and made mostly of gases.

   What does the word *outer* mean?

   A farther away from the sun

   B closer to the sun

   C outside Pluto's orbit

   D made of gas

HINT Use the table on page D18 to help you answer this question.

10. As you move away from the sun, planets —

   A take less time to complete one orbit

   B are smaller than those close to the sun

   C take more time to complete an orbit

   D have no moons

HINT Both the captions and the text on pages D16 and D17 will help you answer this question.

11. The asteroids may be —

   A an exploded planet

   B matter that failed to form a planet

   C alien spacecraft

   D lost moons of Saturn

HINT Captions as well as the main text can provide important information.

12. When does the Northern Hemisphere have the most hours of daylight?

   A on the day of the summer solstice

   B on the day of the autumn equinox

   C on the day of the winter solstice

   D on the day of the spring equinox

HINT What bodies come from beyond Pluto?

13. The solar system bodies with the longest orbits are —

   A planets

   B comets

   C moons

   D asteroids

How Have People Explored the Solar System?

Read pages D22 to D27 in your textbook. Then read each question that follows. Decide which is the best answer to each question. Mark the letter for that answer.

HINT Review the definition of *satellite* to help you answer this question.

14. All of the following are satellites EXCEPT —

   A the moon

   B Earth

   C *Sputnik*

   D the sun

HINT Use the text and timeline on pages D22 and D23 to help you answer this question.

15. Which happened first?

   A Sir Isaac Newton designs a telescope that uses a mirror and lenses.

   B Galileo discovers four of Jupiter's moons.

   C The Maya watch the movements of Venus at Chichen Itza.

   D The phases of the moon are observed.

HINT Only one of these craft orbited only Earth.

16. All of the following craft were sent to gather information about other planets EXCEPT —

   A *Viking I*

   B *Sputnik I*

   C *Voyager 2*

   D *Pioneer*

HINT Carefully read the text on page D26 to help you answer this question.

17. After reading this lesson, what can you conclude about the International Space Station?

   A It is being built in sections.

   B It will have "artificial gravity."

   C The first scientists to occupy the station will be Americans.

   D The station will orbit the moon.

HINT Which of these missions included astronauts?

18. The only mission to return with samples from the place it visited was —

   A *Viking I*

   B the first Space Shuttle mission

   C *Sputnik I*

   D *Apollo*

HINT Reread the lesson looking for important details.

19. This lesson tells you that —

   A plans to build a moon base are being considered

   B construction of a base on the moon has already begun

   C a moon base could be possible within the next twenty years

   D we are still about a century away from building a moon base

Unit D, Chapter 1

Base your answers on the information in this chapter. Read all parts to each question before you begin.

An eclipse occurs when one object passes through the shadow of another. Compare and contrast a lunar eclipse and a solar eclipse. In your discussion, tell how each type of eclipse appears to an observer on Earth.

HINT Which heading introduces information on eclipses?

_____

_____

_____

_____

_____

_____

_____

The moon is the brightest object in the night sky and Earth's nearest neighbor in space, but it can't support life. Use information from the chapter to write a paragraph telling why this is so.

HINT What do all living beings need to survive?

_____

_____

_____

_____

_____

_____

All Suited Up

Space is a cold, dark, airless place. When humans travel in space they must take their environment with them. Space vehicles must contain air. They must be kept at the proper temperature. They must carry water for drinking and washing. People who have to leave the space vehicle to work or explore must wear a space suit. The space suit is another way of taking your environment with you.

Examine the spacesuit on page D25. Think about what it was like to wear this type of suit. Then write a letter to a friend as if you were an Apollo astronaut. Explain what each part of your suit does, and why it is important. Also tell your friend how it feels to wear the suit as you explore the surface of the moon.

Use this page for prewriting or planning activities. Then write your response on a separate sheet of paper.

Writer's Checklist	
IDEAS • Is my message clear? • Do I know enough about my topic? • Have I included interesting details?	**WORD CHOICE** • Will my reader understand my words? • Did I use words I love? • Are my words interesting? • Can I picture it?
ORGANIZATION • Does my paper start out with a bang? • Did I tell things in the best order? • At the end does it feel finished and make you think?	**SENTENCE FLUENCY** • Is my paper easy to read out loud? • Do my sentences begin in different ways? • Are some sentences long and some short?
VOICE • Does this writing really sound like me? • Did I say what I was thinking? • Did I express how I feel?	**CONVENTIONS** • Did I use paragraphs? • Is it easy to read my spelling? • Did I use capital letters in the right place? • Are periods, commas, exclamation marks, and quotation marks in the right places?

What Are the Features of the Sun?

Read pages D38 to D43 in your textbook. Then read each question that follows. Decide which is the best answer to each question. Mark the letter for that answer.

HINT The other words in these sentences provide clues to the meaning of the word *permanent*.

1. Read these sentences from the lesson.

   The sun is Earth's "local star" — the star at the center of the solar system. It has no permanent features, like Earth's mountains and oceans, because the sun is a huge ball of very hot gases.

   What does the word *permanent* mean?

   A very large

   B lasting only for a short time

   C fixed or changeless

   D outstanding

HINT Reread the lesson to answer the question.

2. What happens during photosynthesis?

   A Plants convert the sun's energy into food energy.

   B Water and carbon dioxide are released from tree leaves.

   C Small particles are fused to form larger ones.

   D Energy from the sun travels in waves.

HINT You can find this information under the heading "Exploring the Sun."

3. Where is most of the sun's mass found?

   A in the radiation zone

   B in the convection zone

   C in its core

   D in the photosphere

HINT Read the caption that accompanies the illustration.

4. What does the illustration on the bottom of pages D38–D39 show?

   A the weather patterns on Earth

   B the process of fusion

   C the difference between infrared and ultraviolet waves

   D how the sun's energy travels in waves

HINT Reread the passage following the heading "Solar Features."

5. What might you observe on Earth after a solar flare?

   A The sun would look brighter.

   B A heavy rainstorm might begin.

   C You might observe sunspots.

   D Your compass might not work correctly.

HINT Which details in the passage tell about sunspots?

6. What is one thing that scientists have **NOT** observed about the phenomenon known as sunspots?

   A Sunspots increase and decrease over a period of eleven years.

   B Sunspots appear as bright loops of gas in the sun's corona.

   C Sunspots can produce solar flares.

   D Sunspots appear darker than the rest of the photosphere.

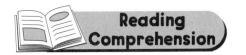
How Are Stars Classified?

Read pages D46 to D51 in your textbook. Then read each question that follows. Decide which is the best answer to each question. Mark the letter for that answer.

HINT The hottest stars are in the blue part of the diagram.

7. After studying the diagram on page D47, you can conclude that —

   A white dwarfs are cooler than red giants

   B the sun is a red giant

   C bright stars are always hot stars

   D stars can be small and very hot

HINT Captions often contain important information.

8. A star reaches the main sequence —

   A after its temperature reaches about 15 million °C

   B after it becomes a red giant

   C just before it becomes a protostar

   D after it uses most of its hydrogen

HINT Review the diagram on pages D48 and D49.

9. Which of the following will happen when the sun becomes a red giant?

   A Earth will be swallowed up by the sun's expanding atmosphere.

   B All of Earth's oceans will freeze.

   C The temperatures on Earth will become too hot for life to exist

   D Venus will become suitable for life.

HINT Prefixes change the meaning of a word.

10. In the word *protostar*, what does the prefix *proto-* mean?

    A heavy

    B coming after

    C imitating

    D giving rise to

HINT What is the main idea of this lesson?

11. What generalization can you make about stars after reading this lesson?

    A All stars change.

    B Most stars are like the sun.

    C All stars produce the same amount of energy.

    D All stars are circled by planets.

HINT Where in the lesson will you find information about observing stars?

12. The telescopes in the Very Large Array detect —

    A visible light using large mirrors

    B radio waves

    C X rays

    D visible light using lenses

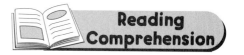
What Are Galaxies?

Read pages D54 to D57 in your textbook. Then read each question that follows. Decide which is the best answer to each question. Mark the letter for that answer.

HINT Under which heading would you find this information?

13. What type of a galaxy is the Milky Way Galaxy?

   A spiral   **C** elliptical

   B barred spiral   **D** irregular

HINT Important vocabulary often is highlighted in yellow.

14. A light-year is a measure of —

   A time

   B speed

   C distance

   D distance in a specific amount of time

HINT Important information is often found at the beginning of a section.

15. The universe is —

   A about 50% dust

   B almost entirely filled with stars

   C mostly empty space

   D 50% dust, 15% stars, 5% planets, 30% empty space

HINT Skim the lesson to find important terms.

16. An example of a galactic cluster is —

   A the Milky Way Galaxy

   B the Local Group

   C the Horsehead Nebula

   D an irregular galaxy

HINT Reread "Types of Galaxies" to find this information.

17. How are elliptical galaxies different from barred and spiral galaxies?

   A Elliptical galaxies do not contain protostars.

   B Barred and spiral galaxies do not form clusters.

   C Elliptical galaxies do not seem to rotate like barred and spiral galaxies.

   D Elliptical galaxies are much bigger than barred and spiral galaxies.

HINT Reread the lesson for specific details.

18. Which statement is **NOT** TRUE?

   A The Virgo Cluster contains more than 100,000 stars.

   B A galactic cluster often appears as a faint smudge in the sky.

   C The gases in a nebula may glow red or green.

   D The "horsehead" in the Horsehead Nebula is actually dust.

HINT Make a generalization about each section as you read.

19. What is the main idea of the section headed "Types of Galaxies"?

   A A spiral galaxy looks like a giant pinwheel spinning through space.

   B About half of all galaxies are elliptical.

   C There are four basic types of galaxies: spiral, barred spiral, elliptical, and irregular.

   D Galaxies contain old and new stars, protostars, dust, and gas.

Unit D, Chapter 2

Base your answers on the information in this chapter. Read all parts to each question before you begin.

The diagram on page D47 is called an H–R diagram, after Hertzsprung and Russell. It is used by astronomers to classify stars. When stars are plotted on the diagram, a trend called the "main sequence" is shown. Explain what is meant by the "main sequence."

HINT Use the illustration on page D47 along with the text on that page to answer this question.

Possible answer: The main sequence is a band of stars that runs from the top left to

the bottom right of the diagram. At the top left of the band are bright, hot, blue stars.

As you move down the band, stars become cooler and less bright. Stars like our sun,

which shines with a yellow-white light, are in the middle of this band. At the bottom

right of the band are cool, dim, red stars. About 95 percent of the stars scientists have

observed are on the main sequence.

Stars may look like they don't change, but they do. Stars go through stages that have been compared to an organism's life cycle. Explain the changes that a typical main sequence star goes through.

HINT The Inside Story on pages D48 and D49 will help you answer this question.

Possible answer: Stars begin within a nebula—a cloud of hydrogen, helium, and dust.

Over millions of years, these particles are attracted to each other by gravity. As they are

squeezed together, they form a protostar. As more material is added to a protostar, its

temperature rises until it begins to glow. After several million years, the temperature of

the center of the protostar reaches 15 million °C, and energy begins to be released. The

star is now on the main sequence. After billions of years, a star's hydrogen begins to

run low. The star begins to expand and becomes a red giant. The red giant's

atmosphere expands a million times and forms a planetary nebula. At the center of the

nebula is a white dwarf, which will shine dimly for billions of years as it cools.

Galaxies

Most of the stars in the universe are in galazies. The universe contains at least a hundred billion galaxies, and each galaxy contains billions of stars. Galaxies, which also contain dust and gas, are classified by their shape. There are four basic galaxy shapes.

spiral   barred spiral   elliptical   irregular

Choose one of these types of galaxies. Use what you have learned in this chapter and information you research to write a description of our galaxy for a classmate. Include a description of its shape, whether it rotates, and what it contains. Include a diagram of your galaxy with your description.

Use this page for prewriting or planning activities. Then write your response on a separate sheet of paper.

<table>
<tr><td colspan="2" align="center">Writer's Checklist</td></tr>
<tr><td>

IDEAS

- Is my message clear?
- Do I know enough about my topic?
- Have I included interesting details?

ORGANIZATION

- Does my paper start out with a bang?
- Did I tell things in the best order?
- At the end does it feel finished and make you think?

VOICE

- Does this writing really sound like me?
- Did I say what I was thinking?
- Did I express how I feel?

</td><td>

WORD CHOICE

- Will my reader understand my words?
- Did I use words I love?
- Are my words interesting?
- Can I picture it?

SENTENCE FLUENCY

- Is my paper easy to read out loud?
- Do my sentences begin in different ways?
- Are some sentences long and some short?

CONVENTIONS

- Did I use paragraphs?
- Is it easy to read my spelling?
- Did I use capital letters in the right place?
- Are periods, commas, exclamation marks, and quotation marks in the right places?

</td></tr>
</table>

Math Practice

Unit D, Chapter 1

Read each question and choose the best answer. Mark the letter for that answer.

1. Which bar graph best represents the data in the table below?

Orbits Around the Sun

Planet	Earth Days
Mercury	88
Venus	225
Earth	365

A

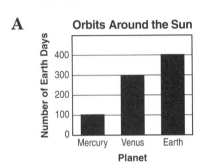

B

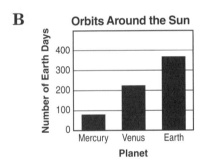

C

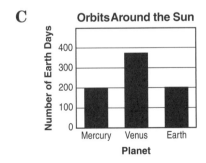

D

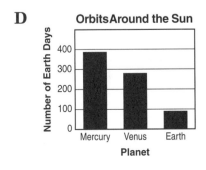

2. Earth rotates at a speed of 1730 kilometers per hour. How many meters per hour does it rotate?

   A 173 meters

   B 1730 meters

   C 17,300 meters

   D 1,730,000 meters

3. The distance to the moon from Earth is best measured in which of the following units?

   A kilometers   **C** centimeters

   B meters   **D** millimeters

4. The largest mare on the moon is 1,248,000 meters across. How many kilometers is this?

   A 12.48 km   **C** 1248 km

   B 124.8 km   **D** 12,480 km

5. Tycho crater on the moon is 87 kilometers across. How many meters is this?

   A 8.7 m   **C** 870 m

   B 87 m   **D** 87,000 m

6. A lunar eclipse lasts around two hours. How many minutes is this?

   A 120 min.   **C** 30 min.

   B 60 min.   **D** 2 min.

Go On

Name _____

Date _____

Math Practice

Unit D, Chapter 1

Read each question and choose the best answer. Mark the letter for that answer.

Use the table below for problems 7 and 8.

Length of Day

Planet	Hours in a day
Jupiter	10
Neptune	16
Earth	24

7. Choose the bar graph that best represents the data in the table.

A

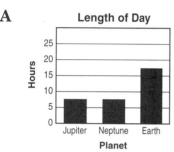

B

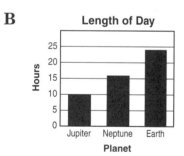

C

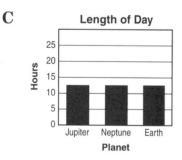

D

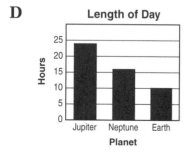

8. Use the table to determine how many more hours are in a day on Earth than on Jupiter.

  A 14 hours   **C** 34 hours

  B 24 hours   **D** 40 hours

9. If one day is 10 hours long on Jupiter, how many hours are in a 7 day week there?

  A 3 hours

  B 17 hours

  C 70 hours

  D 700 hours

10. The expected mass of the space station is best expressed by which of the following units?

  A grams

  B kilograms

  C ounces

  D liters

11. A certain rocket traveled 193 kilometers. How many meters is this?

  A 19.3 m

  B 193 m

  C 19,300 m

  D 193,000 m

Stop

Unit D, Chapter 2

Read each question and choose the best answer. Mark the letter for that answer.

12. The sun's diameter is best measured using which of the following units?

    A millimeters

    B centimeters

    C meters

    D kilometers

13. A sunspot cycle is best measured in which of the following units?

    A years

    B hours

    C minutes

    D seconds

14. The distance from Earth to Mars is 1.5 AU. The distance from Earth to Neptune, is 30.0 AU. How much farther is it to Neptune than Mars?

    A 31.5 AU

    B 28.5 AU

    C 21.5 AU

    D 1.5 AU

15. A rocket travels the 150 million kilometers from Earth to the sun and back. How many AU's did the rocket travel?

    A 150 million

    B less than 1

    C 2

    D 1

16. How many watts brighter is a 60-watt bulb than a 40-watt bulb?

    A 60-watts

    B 40-watts

    C 20-watts

    D 10-watts

17. Ninety-five percent of the stars that scientists have classified are main-sequence stars. What percent are other types of stars?

    A 90%

    B 85%

    C 50%

    D 5%

Go On

Unit D, Chapter 2

Read each question and choose the best answer. Mark the letter for that answer.

18. The Keck telescope has a mirror that measures 10 meters across. How many centimeters is this?

   A   10 cm

   B   100 cm

   C   1,000 cm

   D   10,000 cm

19. Earth's diameter is 12,700 km. The sun's diameter is 1,400,000 kilometers. How many times larger is the sun's diameter than that of Earth's?

   A   over 100 times

   B   over 1,000 times

   C   over 10 times

   D   over 10,000 times

20. A light-year is best measured using which of the following units?

   A   kilometers

   B   inches

   C   hours

   D   minutes

21. A day on Mars is 24 hours, 37 minutes. How many minutes is this?

   A   1640 min.

   B   1477 min.

   C   1440 min.

   D   24.37 min.

22. Choose the type of graph that would best represent the distances from the sun to the planets.

   A   stem-and-leaf plot

   B   circle graph

   C   bar graph

   D   picture graph

23. The lifespan of a star is best measured in —

   A   hundreds of years

   B   thousands of years

   C   millions of years

   D   billions of years

Name _____

Date _____

Unit D, Review

Read each question and choose the best answer. Mark the letter for that answer.

Use the table below for problems 24 and 25.

Orbits Around the Sun

Planet	Earth Days
Jupiter	4,333
Saturn	10,759
Uranus	30,685

24. Choose the bar graph that best represents the data in the table.

A
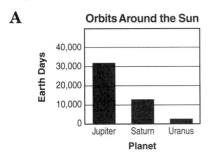

B
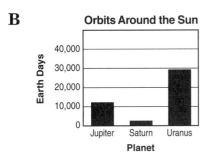

C
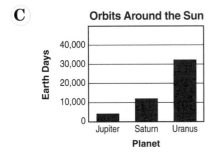

D
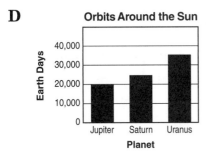

25. Determine how many more days it takes for Uranus to orbit the sun, than Jupiter.

A 26,352 days   **C** 10,759 days

B 19,926 days   **D** 4,333 days

26. Suppose that in a year there are two solar eclipses and five lunar eclipses. How many more lunar eclipses take place than solar eclipses?

A 5   **C** 2

B 3   **D** 1

27. A day on Jupiter is 9 hours and 50 minutes. How many minutes is this?

A 50 min.   **C** 590 min.

B 540 min.   **D** 950 min.

Use the table below for problems 28 and 29.

Earth Hours in a Day

Planet	Earth	Pluto	Mercury
Hours	24	144	1408

28. What type of graph would best represent the data in the table?

A circle graph   **C** stem-and-leaf plot

B picture graph   **D** bar graph

29. Which sentence best describes the data in the table?

A A day on Mercury is longer than a day on Earth.

B A day on Pluto is shorter than a day on Earth.

C A day on Pluto is the same length as on Mercury.

D A day on Earth is longer than a day on Pluto.

How Can Physical Properties Be Used to Identify Matter?

Read pages E6 to E11 in your textbook. Then read each question that follows. Decide which is the best answer to each question. Mark the letter for that answer.

HINT Remember, a physical property of an object can be measured without changing the object into something else.

1. What is **NOT** an example of a physical property?

   A the color of a piece of candy

   B the mass of an automobile

   C how two chemicals combine to form a new chemical

   D the weight of an astronaut on the moon

HINT The caption will help you to understand the illustration's purpose.

2. What is the purpose of the illustration on the upper right-hand corner of page E7?

   A It shows that the foam is heavier than the three weights on the other end of the scale.

   B It shows that the foam and the gel weigh the same amount.

   C It shows that different equipment is used to measure weight and mass.

   D It shows that an object's mass does not change even if the object's shape changes.

HINT Study the diagrams and captions along with the text to answer the question.

3. How is the volume of an irregularly shaped solid measured?

   A by weighing the solid on a spring scale

   B by measuring how much liquid the solid displaces

   C by multiplying the length, width, and height of the solid

   D by dividing the solid's mass by its volume

HINT Reread the passage headed "Density" to find this information.

4. Which physical property would you need to measure or calculate to tell a diamond from a zircon?

   A the mass of the gemstone

   B the volume of the gemstone

   C the density of the gemstone

   D all of the above

HINT Other information in the paragraph that contains this sentence will help you tell the meaning of the word.

5. Read the following sentence from the lesson.

   In a mixture of sugar and water, the sugar dissolves in the water.

   What does the verb *dissolve* mean?

   A to separate into two pure substances

   B to change the physical properties of a substance

   C to become part of a solution

   D to identify whether or not a substance is soluble

Reading Comprehension

How Does Matter Change from One State to Another?

Read pages E14 to E19 in your textbook. Then read each question that follows. Decide which is the best answer to each question. Mark the letter for that answer.

HINT The captions will help you understand how solids, liquids, and gases differ.

6. The particles in a liquid —

  A move more freely than they do in a gas

  B move freely enough to allow a liquid to change its shape

  C are packed close together and move very slowly

  D do not move at all

HINT Use your own knowledge and information on page E17 to answer this question.

7. What can water ice do that dry ice cannot do?

  A melt

  B become a gas

  C change from a solid directly into a gas

  D change its shape

HINT Study The Inside Story on pages E16 and E17 to help you answer this question.

8. Which of the following shows the correct cause and effect?

  A Heat added ⟶ evaporation

  B Heat added ⟶ condensation

  C Heat removed ⟶ evaporation

  D Heat removed ⟶ melting

HINT You will need to understand the entire lesson to answer this question.

9. What does this picture of a glass of lemonade illustrate?

  A freezing and melting

  B boiling and evaporation

  C freezing, condensation, and evaporation

  D melting and condensation

HINT Use the chart on page E18 to answer this question.

10 Which can exist on a planet where the temperature is 2000°C?

  A water oceans

  B lakes of liquid iron

  C water ice caps

  D lakes of liquid carbon

HINT Remember, melting and boiling point are physical properties of a substance.

11. Which of the following statements is **NOT** TRUE?

  A When you change the state of matter you also change the substance.

  B Particles can evaporate from a liquid before the liquid comes to a boil.

  C All substances can change states.

  D Condensation is the opposite of evaporation.

How Does Matter React Chemically?

Read pages E22 to E27 in your textbook. Then read each question that follows. Decide which is the best answer to each question. Mark the letter for that answer.

HINT Remember, chemical reactions form new substances.

12. Which of the following produces a chemical reaction?

    A opening a bottle of soda

    B melting butter

    C burning wood

    D painting a wall

HINT You can use context clues to help you determine the meaning of unknown words.

13. Something that is highly *combustible*—

    A burns easily

    B does not burn

    C conducts electricity

    D absorbs heat

HINT Think about what the girl is trying to show with this experiment.

14. What is the purpose of the balloon in the experiment shown on page E26?

    A to measure the mass of the reactants

    B to show how much vinegar reacted with the baking soda

    C to measure the length of time it took for the reaction to occur

    D to trap the gas produced in the reaction

HINT Read the text on page E24 carefully to look for details.

15. This chart compares iron and rust. It has one mistake. What should the chart say to correct that mistake?

Iron	Shiny	Conducts electricity	Reacts with oxygen	Melts at very high temperatures
Rust	Not shiny	Does not conduct electricity	Does not react with oxygen	Melts at a lower temperature than iron

    A Iron does not conduct electricity.

    B Rust reacts with oxygen.

    C Iron melts at very low temperatures.

    D Rust melts at a higher temperature than iron.

HINT Important vocabulary sometimes appears in italic type.

16. What is an ore?

    A pure metal

    B metal that has other substances combined with it

    C metal in another form, such as a liquid

    D combination of two metals

Unit E, Chapter 1

Base your answers on the information in this chapter. Read all parts to each question before you begin.

You now know that matter can exist in three states. Imagine that you are going to write a report on what makes matter change states. Using examples and details from the lesson, make some notes about a substance and how it can change from a solid to a liquid to a gas.

HINT See The Inside Story for an example of matter as it changes state.

_____

_____

_____

_____

_____

_____

_____

Physical properties are used to identify substances. Explain why boiling points and freezing points can be used to identify substances.

HINT Reread the passage to find the reason boiling and freezing points are good identifiers.

_____

_____

_____

_____

_____

_____

_____

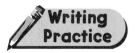

Physical or Chemical Change?

Matter can change in one of two ways: physically or chemically. Examples of physical and chemical changes can be seen around you every day. For example, if you bend a silver spoon, you are changing the shape of the matter that makes up the spoon. This is a physical change. If the spoon becomes tarnished, it is because its silver surface has reacted with the oxygen in the air. This is a chemical change. Scientists sometimes don't agree whether a certain change is physcial or chemical. They use data from experiments to support their positions.

Choose a change that you think is a physical or chemical change, for example, iron rusting, water boiling, or eggs cooking. Write a paragraph that includes experimental data in order to persuade a classmate to support your opinion.

Use this page for prewriting or planning activities. Then write your response on a separate sheet of paper.

Writer's Checklist

IDEAS

- Is my message clear?
- Do I know enough about my topic?
- Have I included interesting details?

ORGANIZATION

- Does my paper start out with a bang?
- Did I tell things in the best order?
- At the end does it feel finished and make you think?

VOICE

- Does this writing really sound like me?
- Did I say what I was thinking?
- Did I express how I feel?

WORD CHOICE

- Will my reader understand my words?
- Did I use words I love?
- Are my words interesting?
- Can I picture it?

SENTENCE FLUENCY

- Is my paper easy to read out loud?
- Do my sentences begin in different ways?
- Are some sentences long and some short?

CONVENTIONS

- Did I use paragraphs?
- Is it easy to read my spelling?
- Did I use capital letters in the right place?
- Are periods, commas, exclamation marks, and quotation marks in the right places?

What Are Atoms and Elements?

Read pages E38 to E43 in your textbook. Then read each question that follows. Decide which is the best answer to each question. Mark the letter for that answer.

HINT This information can be found under the heading "Elements."

1. What is the smallest unit of an element that has all the properties of that element?

   A a proton

   B a neutron

   C an atom

   D an electron

HINT Which heading introduces information about how an atom is made up?

2. What kind of charge does an atom with 20 protons and 20 electrons have?

   A It has a positive charge.

   B It has a negative charge.

   C Its charge depends upon the mass of the nucleus.

   D It has no charge.

HINT Reread the passage headed "The Structure of an Atom."

3. The proton, the electron, and the neutron are all —

   A equally charged

   B subatomic particles

   C larger than the atom

   D in orbit around the nucleus of an atom

HINT In this sentence, the word *corresponds* means "to be equal to."

4. What does an element's atomic number correspond to?

   A to the number of protons in its nucleus

   B to how many atoms it contains

   C to a particular state of matter

   D to another element identified by scientists

HINT Use the table on page E41 to answer this question.

5. What is the second most common element in Earth's crust?

   A oxygen

   B aluminum

   C silicon

   D sodium

HINT These sentences contain context clues that help you to figure out the meaning of the word.

6. Read these two sentences from the lesson.

   Most metals are said to be malleable. They can be hammered or rolled into thin sheets.

   What does the word *malleable* mean?

   A extremely soft

   B capable of being molded without breaking

   C brightly colored

   D very thin

What Are Compounds?

Read pages E46 to E49 in your textbook. Then read each question that follows. Decide which is the best answer to each question. Mark the letter for that answer.

HINT Use context clues to help you determine the meaning of unknown words.

7. What is a metalloid?

   A an artificial element formed by joining two metals

   B a metal that does not conduct electricity

   C an element that has properties of both metals and nonmetals

   D an artificial metal

HINT What does a chemical formula show?

8. This is the chemical formula for a common fertilizer: $(NH_4)_2SO_4$. What elements is it made of?

   A neon, helium, sulfur, oxygen

   B nickel, hydrogen, silicon, osmium

   C nickel, hydrogen, sulfur, oxygen

   D nitrogen, hydrogen, sulfur, oxygen

HINT Use the key on page E46 to help you answer this question.

9. What can you conclude about the elements nitrogen, oxygen, sulfur, and radon?

   A They are all gases.

   B They are all metalloids.

   C They all have the same number of neutrons.

   D They are all nonmetals.

HINT You can find important information by looking for words highlighted in yellow.

10. Which meaning of the word *compound* is the one used in this lesson?

    A a word made up of two words, such as *chalkboard*

    B a substance made up of the atoms of two or more elements

    C a walled-in area containing a group of buildings

    D a sentence having two or more main clauses

HINT Important information is often found at the beginning of a lesson.

11. Who was Dmitri Mendeleev?

    A He manufactured the first artificial element, which was subsequently named after him.

    B He was the first to organize the known elements in a table according to their atomic masses.

    C He restructured the periodic table so that the elements were listed by atomic number.

    D He was the first scientist to rename silicon, boron, arsenic, and antimony as "metalloids."

HINT The formula for table salt is NaCl.

12. Table salt is made of —

    A a metal and a nonmetal

    B two gases

    C two metals

    D a metal and a metalloid

Unit E, Chapter 2

Base your answers on the information in this chapter. Read all parts to each question before you begin.

Elements can be grouped together because they have similar properties. How are elements grouped in the periodic table?

HINT Study the periodic table on pages E46–E47 to find additional details.

_____

_____

_____

_____

_____

_____

_____

_____

You know that all matter is made of atoms. Some substances are made of just one type of atom. Others are made of many different atoms. What happens when atoms form a compound? Include an example of this type of chemical change in your explanation.

HINT Which heading introduces this information?

_____

_____

_____

_____

_____

_____

Describing Elements

In ancient times, people thought there were only four elements: air, water, earth, and fire. As the science of chemistry grew, people discovered that fire was not an element at all, and that air, water, and earth were themselves made of elements. Eventually over 100 elements of matter were discovered. Scientists arranged these elements by their properties on a periodic table.

Examine the table on pages E46 and E47, and then choose an element. Research the element's properties and uses. Then write a description of that element for a classroom journal.

Use this page for prewriting or planning activities. Then write your response on a separate sheet of paper.

Writer's Checklist

IDEAS

- Is my message clear?
- Do I know enough about my topic?
- Have I included interesting details?

ORGANIZATION

- Does my paper start out with a bang?
- Did I tell things in the best order?
- At the end does it feel finished and make you think?

VOICE

- Does this writing really sound like me?
- Did I say what I was thinking?
- Did I express how I feel?

WORD CHOICE

- Will my reader understand my words?
- Did I use words I love?
- Are my words interesting?
- Can I picture it?

SENTENCE FLUENCY

- Is my paper easy to read out loud?
- Do my sentences begin in different ways?
- Are some sentences long and some short?

CONVENTIONS

- Did I use paragraphs?
- Is it easy to read my spelling?
- Did I use capital letters in the right place?
- Are periods, commas, exclamation marks, and quotation marks in the right places?

Name _____

Date _____

Unit E, Chapter 1

Read each question and choose the best answer. Mark the letter for that answer.

1. Choose the bar graph that best represents the data in the table at the right.

Hot and Cold Temperatures

Temperatures	Water Freezes	Human Body	Water Boils
°C	0	37	100

A
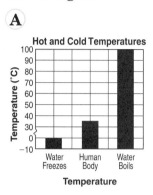

B
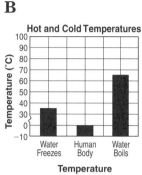

C
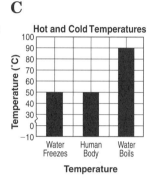

D
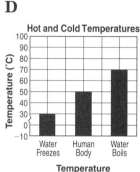

2. Find the volume of the box at the right.

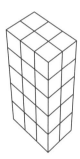

A 10 cubic units

B 16 cubic units

C 30 cubic units

D 352 cubic units

3. Density = mass ÷ volume. If the mass of a copper marble is 89.6 grams and its volume is 10 cubic centimeters, what is its density?

A 0.896 g/cc   **C** 89.6 g/cc

B 8.96 g/cc   **D** 896 g/cc

4. The density of a zircon is 4.7 grams per cubic centimeter. The density of a diamond is 3.5 grams per cubic centimeter. How much more dense is the zircon than the diamond?

A 12.12 grams per cubic centimeter

B 8.2 grams per cubic centimeter

C 4.7 grams per cubic centimeter

D 1.2 grams per cubic centimeter

Use the circle graph below for problems 5 and 6.

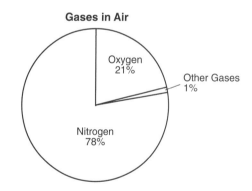

5. Air contains the most of which gas?

A oxygen

B nitrogen

C other gas

D cannot be determined

6. According to the circle graph, what percent of air is oxygen?

A 1 %   **C** 78 %

B 21 %   **D** 100 %

Unit E, Chapter 1

Read each question and choose the best answer. Mark the letter for that answer.

Use the table below for problems 7 and 8.

Melting Points

Substance	°C
Water	0
Table Salt	801
Iron	1535

7. Choose the bar graph that best represents the data in the table.

A
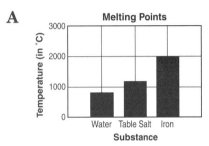

B
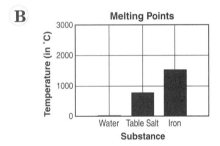

C
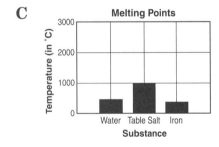

D
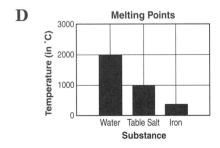

8. Determine the difference in the melting points for table salt and iron.

A 1535 °C   **C** 801 °C

B 1000 °C   **D** 734 °C

9. The volume of a gas varies with pressure. The higher the pressure, the smaller the volume. If a sample of gas has a volume of 150 milliliters, which will be true if the pressure increases?

A The volume will be less than 150 mL.

B The volume will be more than 150 mL.

C The volume will stay 150 mL.

D cannot be determined

10. What is the volume of the gas that fills the box at the right?

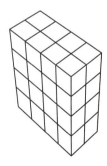

A 10 cubic units

B 16 cubic units

C 32 cubic units

D 442 cubic units

11. The melting point for carbon is 3550°C; its boiling point is 4880°C. What is the difference in degrees Celsius between the boiling point and melting point of carbon?

A 3550°C

B 1330°C

C 1000°C

D 50°C

Unit E, Chapter 1

Read each question and choose the best answer. Mark the letter for that answer.

12. If you pour 50.4 milliliters of oil, 30.1 milliliters of water and 40.9 milliliters of alcohol into one container, how many milliliters will you have?

   A 121.4 mL

   B 80.5 mL

   C 71 mL

   D 40.9 mL

13. If you have 3000 milliliters of water, how many liters do you have?

   A 3000 L

   B 300 L

   C 30 L

   D 3 L

14. What type of graph would best represent the percent of different elements that make up water?

   A line graph

   B circle graph

   C bar graph

   D stem-and-leaf plot

15. Mass is best measured in which of the following units?

   A liters

   B meters

   C grams

   D degrees Celsius

16. A plant and a rock are placed on the balance below. Which of the following statements is TRUE?

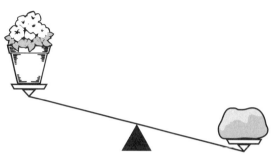

   A The plant has more mass than the rock.

   B The rock has more mass than the plant.

   C The plant and rock have the same mass.

   D The rock's mass is 10.5 grams.

17. Which instrument would you use to measure the volume of a liquid?

   A balance

   B meterstick

   C graduated cylinder

   D none of these

Name _____

Date _____

Unit E, Chapter 2

Read each question and choose the best answer. Mark the letter for that answer.

Use the table below for problems 18 and 19.

Types of Elements

Type	Number of Elements
Metals	88
Nonmetals	17
Metaloids	7

18. Which type of graph would best represent the data in the table?

  A bar graph

  B line graph

  C circle graph

  D stem-and-leaf plot

19. According to the table above, which of the following statements is reasonable?

  A There are more nonmetals than metals.

  B There are more metals than nonmetals and metaloids combined.

  C There are more metaloids than metals and nonmetals.

  D Nonmetals are more useful than nonmetals.

20. What tool would best measure the weight of a small piece of iron?

  A graduated cylinder

  B spring scale

  C meterstick

  D balance

21. What tool would best measure the length of a piece of copper?

  A graduated cylinder

  B spring scale

  C meterstick

  D balance

22. In 1912, the proton was discovered. In 1932, the neutron was discovered. How many years was it between the time the proton was discovered and the time the neutron was discovered?

  A 12 years

  B 20 years

  C 32 years

  D 1920 years

23. Oxygen has 12 electrons. Iron has 31 electrons. How many more electrons does iron have than oxygen?

  A 43

  B 31

  C 19

  D 12

24. How many oxygen atoms are in each molecule of glucose — $C_6H_{12}O_6$?

  A 0

  B 12

  C 24

  D 6

Math Practice

Unit E, Review

Read each question and choose the best answer. Mark the letter for that answer.

Use the table at the right for problems 25–27.

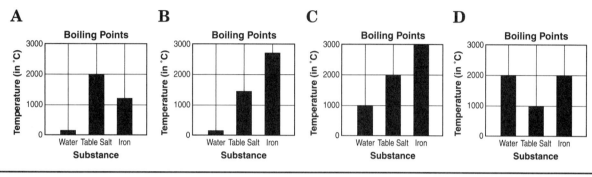

Boiling Points

Substance	Water	Table Salt	Iron
°C	100	1413	2750

25. Choose the line graph that best represents the data in the table.

A   **B**   **C**   **D**

26. How many degrees Celsius different are the boiling points for iron and table salt?

  A   3000°C   **C**   1413°C

  B   1337°C   **D**   1000°C

27. According to the table, which statement would be a reasonable outcome of a race to get a substance to boil?

  A   The person trying to boil the iron would win.

  B   The person trying to boil the table salt would win.

  C   The person trying to boil the water would lose.

  D   The person trying to boil the water would win.

28. In the picture at the right, what is the measure of the liquid?

  A   50 mL

  B   100 mL

  C   125 mL

  D   150 mL

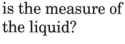

29. Find the volume of the box.

  A   30 cubic units

  B   16 cubic units

  C   10 cubic units

  D   7 cubic units

30. A marble and a coin are placed on the balance below. Which of the following statements is TRUE?

  A   The coin has more mass.

  B   The marble has more mass.

  C   Their masses are the same.

  D   The marble has a mass of 10 g.

What Forces Affect Objects On Earth Every Day?

Read pages F6 to F9 in your textbook. Then read each question that follows. Decide which is the best answer to each question. Mark the letter for that answer.

HINT The text and captions on page F6 will help you answer this question.

1. Brittany skated from her house to the supermarket. During which part of her trip did her inline skates encounter the LEAST amount of friction?

   A on the loose gravel of her driveway

   B through the vacant lot

   C on the new pavement between Cedar and Crosby Streets

   D through the dry leaves in the supermarket's parking lot

HINT What does the strength of gravitation depend on?

2. Which of Jupiter's inner moons is MOST affected by Jupiter's gravity?

HINT Use context clues to determine the meaning of unfamiliar words.

3. The word *repel* means —

   A turn   **C** encourage

   B push away   **D** enlarge

HINT Important information often is found at the beginning of a section.

4. What is the main idea of the section headed "Forces"?

   A The three forces that affect objects on Earth are friction, magnetism, and gravity.

   B The force of a magnetic field attracts objects to a magnet.

   C A force is any push or pull that causes an object to move, stop, or change direction.

   D Forces can act directly or at a distance.

HINT Reread page F8 to help you answer this question.

5. In which pair of objects is the gravitational attraction equal?

   A
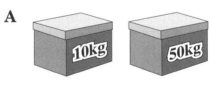

   B
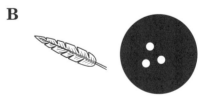

   C

   D

What Are Balanced and Unbalanced Forces?

Read pages F12 to F15 in your textbook. Then read each question that follows. Decide which is the best answer to each question. Mark the letter for that answer.

HINT Remember, it is often a good idea to look for key ideas in the first passage of a lesson.

6. What are balanced forces?

   A forces that have the same weight and size

   B forces that are used at the same time

   C forces that are equal in size and opposite in direction

   D forces that produce friction

HINT Context clues in the sentence will help you determine the meaning of the word.

7. Read this sentence from the lesson.

   Suppose you exert a force by pushing on a very heavy object, such as a sturdy wall.

   What does the word *exert* mean?

   A to cancel

   B to put forth

   C to stop

   D to respond to

HINT The passage contains examples of different effects caused by unbalanced forces.

8. What happens when one force on an object is greater than another force?

   A The object's motion changes.

   B The two forces cancel each other out.

   C The object always moves forward.

   D The object always stops moving.

HINT The author includes this example under the heading "Balanced and Unbalanced Forces."

9. When a cyclist is going downhill, what makes the bicycle speed up?

   A the force of friction

   B two balanced forces

   C the force of the ground

   D the unbalanced force of gravity

HINT Study the passage headed "Net Force" to figure out the answer.

10. If a stone weighs 75 newtons and you apply a force of 150 newtons to lift it, the net force acting on the stone is

   A 225 newtons, up

   B 75 newtons, up

   C 225 newtons, down

   D 75 newtons, down

HINT Use the words highlighted in yellow to help you locate this information.

11. What causes an object to accelerate?

   A Balanced forces act upon the object.

   B The force of gravity acts on the object.

   C A pair of forces act on the object.

   D Unbalanced forces act upon the object.

What Is Work and How Is It Measured?

Read pages F18 to F23 in your textbook. Then read each question that follows. Decide which is the best answer to each question. Mark the letter for that answer.

HINT Use the highlighted terms to locate important information.

12. What is the main idea of the section headed "Work and Effort"?

   A Work and effort are not necessarily the same thing.

   B In scientific terms, work is the use of a force to move an object through a distance.

   C Without the right tools and planning, a gardener can exert a lot of effort but not do any work.

   D Activities that seem fun would be classified as work by scientists.

HINT Important information is often found at the beginning of a section.

13. According to this lesson, what is a machine?

   A any device that runs on electricity, solar power, or batteries

   B something that makes work seem easier by changing the size or direction of a force

   C the simplest parts into which a device can be broken down

   D anything that allows a person to do work without exerting effort

HINT Important information is sometimes found in captions.

14. The washer in a faucet is —

   A a wedge

   B a screw

   C a prop

   D a wheel

HINT Use details from the lesson to help you formulate the lesson's main idea.

15. What do a pulley, lever, inclined plane, wedge, screw, and wheel and axle have in common?

   A They are present in all machines.

   B They all turn or cause objects to turn.

   C They are all machines.

   D They all require the same force to run.

HINT What is one thing a machine does **NOT** do?

16. Which of the following is **NOT** a FACT?

   A A lever allows a person to exert a larger force over a shorter distance.

   B Pulleys help reduce the amount of work to be done.

   C A wheel and axle can change a twisting force into a downward force.

   D A screw can change a force's direction.

HINT Important words sometimes are indicated in italic type.

17. What is a joule?

   A a unit of length

   B a unit of work

   C a size measurement

   D a unit of time

Unit F, Chapter 1

Base your answers on the information in this chapter. Read all parts to each question before you begin.

Sometimes the everyday meanings of words are different from their scientific definitions. Give the scientific definition for work and power. Then provide an example for each.

HINT You can find this information in the first two passages of the lesson.

_____

_____

_____

_____

_____

_____

Every time you ride your bike or turn on a faucet, you are using machines. How do machines make work easier for people? In your explanation include at least one example of machines making work easier.

HINT Which heading introduces this information?

_____

_____

_____

_____

_____

_____

_____

Explaining Sayings About Forces

In this chapter you learned about some of the many forces that are part of everyday life. You learned that if all the forces acting on an object are balanced, the object does not move. Unbalanced forces result in motion. Although we may not think about the forces acting together in our daily lives, popular phrases in our language refer to those same forces. Consider, for example, the phrase "caught between a rock and a hard place." Or, "An unstoppable force meets an immovable object."

Both of these figures of speech refer to forces. What does each mean? Write one or two paragraphs for a classmate to explain what you think these phrases mean. Include a drawing with your explanation to help make your meaning clear.

Use this page for prewriting or planning activities. Then write your response on a separate sheet of paper.

Writer's Checklist

IDEAS
- Is my message clear?
- Do I know enough about my topic?
- Have I included interesting details?

ORGANIZATION
- Does my paper start out with a bang?
- Did I tell things in the best order?
- At the end does it feel finished and make you think?

VOICE
- Does this writing really sound like me?
- Did I say what I was thinking?
- Did I express how I feel?

WORD CHOICE
- Will my reader understand my words?
- Did I use words I love?
- Are my words interesting?
- Can I picture it?

SENTENCE FLUENCY
- Is my paper easy to read out loud?
- Do my sentences begin in different ways?
- Are some sentences long and some short?

CONVENTIONS
- Did I use paragraphs?
- Is it easy to read my spelling?
- Did I use capital letters in the right place?
- Are periods, commas, exclamation marks, and quotation marks in the right places?

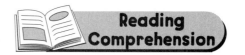
How Are Motion and Speed Related?

Read pages F34 to F37 in your textbook. Then read each question that follows. Decide which is the best answer to each question. Mark the letter for that answer.

HINT Under which heading would you find this information?

1. What is the difference between speed and velocity?

   A Velocity is the average speed of a moving object over time.

   B An object can speed up, but its velocity is always constant.

   C Velocity is speed in a particular direction.

   D Velocity is the scientific term for speed.

HINT Which object would have the most momentum?

2. Which is the hardest to stop?

   A a loaded moving van traveling at 75 miles an hour

   B a snowmobile traveling at 75 miles an hour

   C a car traveling at 70 miles an hour

   D a loaded, 100-car freight train traveling at 7 miles an hour

HINT Look for important words that are highlighted in yellow.

3. In this lesson, the word *momentum* means —

   A a person's frame of reference

   B the amount of force a moving object has

   C a change in velocity

   D the average speed an object moves

HINT Skim the text to locate important words and details.

4. Any change in velocity is called —

   A speed   **C** position

   B momentum   **D** acceleration

HINT The table on page F34 shows examples of how to solve this type of problem.

5. How should you figure out the answer to this problem?

   Sparky raced Fred to the oak tree. The distance was 604 cm. Sparky made it in 12 seconds. What was his average speed?

   A Multiply Sparky's speed by the number of seconds it took Sparky to reach the bush.

   B Divide the distance to the bush by the number of seconds it took Sparky to reach it.

   C Multiply the distance to the bush by the amount of time it took Sparky to reach it.

   D Divide the number of seconds it took Sparky to reach the bush by the distance to the bush.

HINT Reread the material under "Acceleration" to answer this question.

6. When are you **NOT** accelerating?

   A when you are slowing down

   B when you turn a corner

   C when you stop

   D when you are moving at a steady speed in one direction

What Are the Three Laws of Motion?

Read pages F40 to F45 in your textbook. Then read each question that follows. Decide which is the best answer to each question. Mark the letter for that answer.

HINT Reread the first passage of the lesson to summarize key ideas.

7. According to the lesson, what was Newton the first scientist to discover?

   A the position of Earth in the solar system

   B that everything is made of four basic elements

   C that the movements of objects could be described in terms of forces

   D the physical properties of matter

HINT How would you summarize what you have learned about Newton's three laws?

8. What do the laws of motion apply to?

   A only objects on planet Earth

   B only objects that are in motion

   C only nonliving objects

   D objects on Earth and elsewhere in the solar system

HINT The captions of illustrations contain important details.

9. In space, a spacecraft will continue traveling at the same speed for an indefinite amount of time. This is an example of —

   A conservation of momentum

   B the first law of motion

   C reaction force

   D the second law of motion

HINT Other words in these sentences will help you figure out the meaning of the word.

10. Read these sentences from the lesson.

    Because of its great mass, this train has a lot of inertia. It takes a long time to get it moving and a long time to stop it.

    What does the word *inertia* mean?

    A the tendency of a body to resist any change in its motion

    B size and weight

    C ability to use force

    D capacity for travel at high speeds

HINT Scan the lesson to locate these examples and see which laws each one illustrates.

11. What is an example of conservation of momentum?

    A In a traffic collision, the car that is hit from behind moves while the car that hit it stops.

    B A person pushing against a solid wall cannot move the wall.

    C An apple thrown into the air will fall back to Earth.

    D A light rocket needs a smaller engine to lift it than a heavier rocket needs.

Why Do Planets Stay in Orbit?

Read pages F48 to F51 in your textbook. Then read each question that follows. Decide which is the best answer to each question. Mark the letter for that answer.

HINT Reread for important ideas to answer this question.

12. An object that is launched will fall back to Earth when —

A its gravity is greater than its inertia

B its inertia is greater than its gravity

C its speed is greater than Earth's gravity

D its gravity and its inertia are equal

HINT Review page F48 to help you answer this question.

13. The first law of motion is illustrated by —

A Earth's orbit around the sun

B a satellite's orbit around a planet

C both A and B

D Earth's rotation on its axis

HINT The caption on page F51 will help you answer this question.

14. Which statement is TRUE?

A Newton thought the universe was a giant machine.

B Newton built a giant machine to represent the universe.

C Newton pictured the universe as operating like a giant machine.

D Newton's Laws showed that the universe was not a giant machine.

HINT Review the laws of universal gravitation.

15. Where is gravity absent?

A

B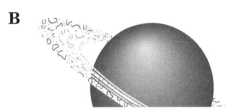

C

D Gravity is always present.

HINT Compare this diagram to the one on pages F48 and F49.

16. How should you label this diagram?

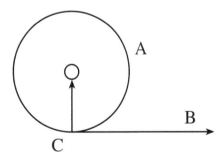

A A is inertia, B is gravity, C is the orbit

B A is gravity, B is the orbit, C is inertia

C A is the orbit, B is gravity, C is inertia

D A is the orbit, B is inertia, C is gravity

Unit F, Chapter 2

Base your answers on the information in this chapter. Read all parts to each question before you begin.

You have learned that objects will stay in motion unless acted upon by an outside force. In the activity, you saw how a washer would fly away because of its motion. What keeps the moon in orbit around Earth? Include information about the forces acting on the moon in your answer.

HINT Use the illustration on page F50 to answer this question.

_____

_____

_____

_____

_____

Isaac Newton stated the three laws of motion. He also stated the law of universal gravitation. Explain what this law states, and give several examples of this law at work in our solar system.

HINT Reread the text on page F49 to help you answer this question.

_____

_____

_____

_____

_____

_____

_____

Name _____

Date _____

Speed and Velocity

One way to describe the motion of an object is to describe its speed. Speed is the measure of an object's motion in a given amount of time. In this chapter you learned that speed and velocity are related, but they are not the same thing.

Suppose you are planning to give a demonstration of speed and velocity to a class of third graders using two toy cars. You will also show your audience the formula for finding the average speed of each car. Explain how you will perform this demonstration. Use information from the lesson to help you prepare.

Use this page for prewriting or planning activities. Then write your response on a separate sheet of paper.

Writer's Checklist

IDEAS

- Is my message clear?
- Do I know enough about my topic?
- Have I included interesting details?

ORGANIZATION

- Does my paper start out with a bang?
- Did I tell things in the best order?
- At the end does it feel finished and make you think?

VOICE

- Does this writing really sound like me?
- Did I say what I was thinking?
- Did I express how I feel?

WORD CHOICE

- Will my reader understand my words?
- Did I use words I love?
- Are my words interesting?
- Can I picture it?

SENTENCE FLUENCY

- Is my paper easy to read out loud?
- Do my sentences begin in different ways?
- Are some sentences long and some short?

CONVENTIONS

- Did I use paragraphs?
- Is it easy to read my spelling?
- Did I use capital letters in the right place?
- Are periods, commas, exclamation marks, and quotation marks in the right places?

What Are Kinetic and Potential Energy?

Read pages F62 to F65 in your textbook. Then read each question that follows. Decide which is the best answer to each question. Mark the letter for that answer.

HINT Review the vocabulary on page F62.

1. Which illustrates potential energy?

   A a rocket sitting on the launch pad
   B the engines firing
   C the rocket lifting off
   D the astronauts waving

HINT Use context clues to help answer this question.

2. In this lesson, the word *transformation* means —

   A carrying across
   B spreading out
   C change
   D rising to another level

HINT Important information sometimes appears in italic type

3. What is the law of conservation of energy?

   A All things must have either potential or kinetic energy.
   B All matter can be transformed into energy.
   C Energy can neither be created nor destroyed.
   D All energy that is added to a system must be used or stored.

HINT Under what heading would you find this information?

4. What is the difference between kinetic energy and potential energy?

   A An object has potential energy because of its position or condition. It has kinetic energy when it is moving.
   B An object can go from having potential energy to having kinetic energy but not the reverse.
   C Living things have kinetic energy. Only nonliving things such as books can have potential energy.
   D An object has potential energy for a fraction of a second, just before it is set in motion. Then it has kinetic energy.

HINT Review lesson vocabulary to help you answer this question.

5. Which is labeled INCORRECTLY?

   A Chemical Energy
   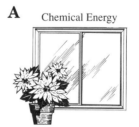

   C Elastic Potential Energy
   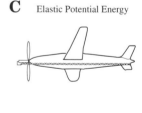

   B Gravitational Potential Energy

   D Thermal Energy
   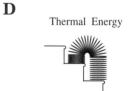

What Is Electric Energy?

Read pages F68 to F73 in your textbook. Then read each question that follows. Decide which is the best answer to each question. Mark the letter for that answer.

HINT Under which heading will you find this information?

6. Which has a negative charge?

   A $+ + + + - - - -$

   B $+ + + + + - - - - - -$

   C $+ + + + + + - - - - -$

   D $+ -$

HINT Review the captions on page F70 and F71 to help you answer this question.

7. What can you tell about this set of lights?

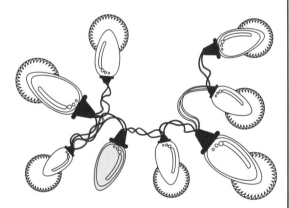

   A The current is not flowing to the burned out bulb.

   B The bulbs are on a series circuit.

   C Something is wrong with the entire set of lights and even replacing the bulb won't help.

   D The bulbs are on a parallel circuit.

HINT Review vocabulary highlighted in yellow.

8. What is an electromagnet?

   A an electric field that attracts objects, similar to the way a magnet attracts metal objects

   B a coil of current-conducting wire wrapped around an iron bar

   C a very large magnet

   D a magnetic field that is produced when static electricity builds up in the atmosphere

HINT Reread the section "Electric Force" to help you answer this question.

9. Electrons flow —

   A from the atmosphere to the ground

   B in circular paths

   C from negatively charged objects to electrically neutral objects

   D from negatively charged objects to positively charged objects

HINT Important information is often contained in captions.

10. Lightning illustrates —

   A a build-up of electrons in the air

   B a build-up of heat

   C heat flow

   D how the ground can become a temporary electromagnet

What Are Light and Sound Energy?

Read pages F76 to F81 in your textbook. Then read each question that follows. Decide which is the best answer to each question. Mark the letter for that answer.

HINT Review the section "Light Energy" to answer this question.

11. Which statement about light is **NOT TRUE**?

    A Light travels in a straight line until it strikes something.

    B A red rose reflects red light, and its green leaves reflect green light.

    C When light strikes water, it bends.

    D A mirror absorbs light.

HINT Under which heading would you find infomation about lenses?

12. A convex lens makes —

    A all objects appear brighter

    B distant objects appear closer

    C nearby objects appear larger

    D nearby objects appear smaller and further away

HINT Important information is sometimes found in captions.

13. Which part of the eye is blue in a blue-eyed person?

    A iris   **C** retina

    B pupil   **D** lens

HINT Which part of the eye bends light?

14. Corrective lenses such as contacts are most similar to the eye's —

    A iris   **C** retina

    B cornea   **D** pupil

HINT Review the text and illustrations on page F79 to help you answer this question.

15. What causes the eardrum to vibrate?

    A vibrating hair cells in the cochlea

    B a change in air pressure

    C movement of the hammer, anvil, and stirrup

    D movement of the fluid in the cochlea

HINT Study the data in the table on page F80 to answer this question.

16. Sound travels through —

    A air slower than it travels through water

    B steel faster than it travels through water

    C water slower than it travels through air

    D a vacuum faster than it travels through air

HINT Review the meaning of frequency before you answer this question.

17. High-pitched sounds —

    A do not travel in a straight line

    B travel fastest in a vacuum

    C travel faster than low-pitched sounds

    D are not detected by the human ear

What Are Thermal and Chemical Energy?

Read pages F84 to F87 in your textbook. Then read each question that follows. Decide which is the best answer to each question. Mark the letter for that answer.

HINT Review the information under the heading "Chemical Energy" to answer this question.

18. Which of the following is **NOT** a FACT?

   A Chemical energy can be released as several forms of kinetic energy.

   B Some chemical reactions give off energy.

   C Some chemical reactions take in energy.

   D Body processes have no effect on thermal energy.

HINT To answer this question, make a generalization from the table on page F86.

19. Which of the following statements is TRUE?

   A Snacks are unhealthy and high in Calories.

   B Fruits are lower in Calories than dairy products.

   C A healthy breakfast is under 400 Calories.

   D Foods that provide the most potential energy are the healthiest.

HINT Understanding word parts can help you interpret new vocabulary.

20. The word *thermal* comes from the Greek word *therm* meaning —

   A chemical   **C** movement

   B heat   **D** molecule

HINT Reread page F85 to help you answer this question.

21. What is the main idea of the section headed "Transferring Thermal Energy"?

   A Only conduction and convection can transfer heat through matter.

   B Radiation is the transfer of thermal energy by electro-magnetic waves.

   C Thermal energy can be transferred by conduction, convection, or radiation.

   D The movement of hot water rising, cooling, sinking, being reheated, and then rising again is an example of convection.

HINT Words highlighted in yellow are important vocabulary terms.

22. What is temperature?

   A the amount of heat that a warm substance can transfer to a cool one

   B the number of molecules in a substance

   C the average kinetic energy of all the molecules in an object

   D the average number of collisions between molecules in a substance

Unit F, Chapter 3

Base your answers on the information in this chapter. Read all parts to each question before you begin.

Lenses bend light rays. Compare and contrast convex and concave lenses. Explain how each type is used.

HINT You can find several examples of how lenses are used under the heading "Lenses" on page F77.

_____

_____

_____

_____

_____

_____

_____

Your teacher calls for attention, and you turn toward him or her and get ready to listen. How did the sound energy travel from your teacher to you?

HINT Remember to use your own words and to include the most important points in your summary.

_____

_____

_____

_____

_____

_____

_____

_____

Changing Chemical Energy

In this chapter you learned that, although there are many forms of energy, there are only two basic forms: kinetic and potential. You also learned that energy cannot be created or destroyed, but can be changed from one form to another. Chemical energy is a form of potential energy. Mechanical energy is a form of kinetic energy.

Write an article for your school newspaper explaining how the body changes chemical energy into mechanical energy. Make your explanation easier for your classmates to understand by including examples of how the body uses Calories.

Use this page for prewriting or planning activities. Then write your response on a separate sheet of paper.

Writer's Checklist

IDEAS
- Is my message clear?
- Do I know enough about my topic?
- Have I included interesting details?

ORGANIZATION
- Does my paper start out with a bang?
- Did I tell things in the best order?
- At the end does it feel finished and make you think?

VOICE
- Does this writing really sound like me?
- Did I say what I was thinking?
- Did I express how I feel?

WORD CHOICE
- Will my reader understand my words?
- Did I use words I love?
- Are my words interesting?
- Can I picture it?

SENTENCE FLUENCY
- Is my paper easy to read out loud?
- Do my sentences begin in different ways?
- Are some sentences long and some short?

CONVENTIONS
- Did I use paragraphs?
- Is it easy to read my spelling?
- Did I use capital letters in the right place?
- Are periods, commas, exclamation marks, and quotation marks in the right places?

How Do People Use Fossil Fuels?

Read pages F98 to F101 in your textbook. Then read each question that follows. Decide which is the best answer to each question. Mark the letter for that answer.

HINT Use the caption and table on F98 to find the meaning of the word.

1. Read this sentence from the lesson.

   When fossil fuels are taken from the ground, their chemical energy can be converted to thermal energy by burning.

   What does the word *converted* mean?

   A heated   **C** joined

   B used by human beings   **D** changed into another form

HINT Reread the passage headed "Using Fossil Fuels" to find the answer.

2. Which type of fuel is most often used in the United States to generate electricity?

   A hardwood   **C** coal

   B natural gas   **D** petroleum

HINT How long ago did fossils form?

3. Why are fossil fuels considered "nonrenewable"?

   A They are extremely expensive to use.

   B They take millions of years to form.

   C They are the main source of energy for industrial nations.

   D They do not all release the same amount of heat.

HINT How does the author organize reasons for using alternative sources of energy? Which reasons seem strongest to you?

4. According to the author, why should we use sources of energy other than fossil fuels?

   A Burning fossil fuels releases large amounts of carbon dioxide into the air.

   B Fossil fuels do not give off much heat.

   C Burning hardwood is more practical than burning fossil fuels.

   D Geologists are continually finding new fossil fuel deposits.

HINT What is the author's reason for presenting examples of alternative energy sources?

5. In the passage "Alternatives to Fossil Fuels," what is the author's main purpose?

   A to explain why fossil fuels are still the best energy sources

   B to describe Californians as leaders in the energy field

   C to describe a variety of energy sources

   D to tell about personal experiences using energy alternatives

HINT Which suggestion seems most reasonable to you?

6. What does the author suggest as one way to reduce the use of fossil fuels?

   A stop using fossil fuels entirely

   B recycle materials such as plastics

   C go back to burning wood for fuel

   D convince utility companies to convert to wind energy

How Can Moving Water Generate Electricity?

Read pages F104 to F107 in your textbook. Then read each question that follows. Decide which is the best answer to each question. Mark the letter for that answer.

HINT Review pages F104 and F105 to answer this question.

7. How is water used to generate electricity?

   A The falling water spins the blades of a turbine, and then the rotating blades spin the shaft of an electric generator.

   B Oxygen is separated from the water, cooled to a liquid state, and used as fuel to run turbines.

   C Water flows through turbines, setting their blades in motion. The motion of the blades sets up an electric field.

   D Water is passed over metal plates in the presence of a magnetic field, cooled, sent through a turbine, and converted to electrical current.

HINT Use context clues to help you figure out the meaning of unknown words.

8. What is an inexhaustible source of energy?

   A one that can be made to operate at constant power

   B an energy source, such as coal, that comes from deep inside Earth

   C an energy source that will never run out

   D a "clean" energy source that does not pollute the environment

HINT The photographs in the first part of the lesson can help you answer this question.

9. What is a fish ladder?

   A a nickname for a dam with a smaller dam behind it

   B a small river that circles a dam so that fish can swim past the dam

   C an opening in the dam that acts as a waterfall

   D a long series of small steps over which water flows so fish can get past a dam

HINT Important vocabulary is sometimes indicated by italic type.

10. What causes the supertides at the Bay of Fundy?

   A the violent weather that often occurs along the rocky coast of Nova Scotia

   B the uneven coastline

   C the narrow bay, which causes the water level to change dramatically

   D the extreme latitude of the Bay of Fundy

What Other Sources of Energy Do People Use?

Read pages F110 to F113 in your textbook. Then read each question that follows. Decide which is the best answer to each question. Mark the letter for that answer.

HINT Important vocabulary is highlighted in yellow.

11. Which of the following would be considered biomass?

  A coal   **C** a pile of leaves

  B natural gas   **D** a mound of dirt

HINT Read the labels on graphs carefully.

12. What can you conclude from the information given in the circle graph on page F112?

  A Most of the world's energy comes from nonpolluting sources.

  B Geothermal, solar, and wind power are becoming more and more popular.

  C Nuclear energy is steadily falling out of favor.

  D Most of our energy is produced by burning fossil fuels.

HINT Look for the subsection "Nuclear Energy."

13. Energy is produced from nuclear fission by —

  A splitting the nuclei of atoms

  B stripping the electrons from atoms

  C forcing atoms to join

  D enlarging the nuclei of atoms by adding additional protons

HINT Reread the text and captions on page F111 to help you answer this question.

14. What is a wind farm?

  A an area such as a meadow or an open plain where it is usually very windy

  B a place where windmills are manufactured

  C an area where many windmills produce large amounts of electricity

  D a place where the buildings are positioned to produce strong wind currents which are then used to generate electricity

HINT Summarize each section as you read the lesson.

15. Which of the following generalizations can you make based on the information in this lesson?

  A There seems little chance that people will be able to develop alternative sources of energy.

  B There are many alternatives to fossil fuels as energy sources.

  C Any energy source that is nonpolluting is also inefficient.

  D Our energy sources can no longer keep up with our energy demands.

HINT Under which section will you find this information?

16. Which of the following potential energy sources is presently in the experimental stage?

  A tidal energy

  B fission

  C ocean thermal energy conversion

  D using hydrogen that has been separated from ocean water

Unit F, Chapter 4

Base your answers on the information in this chapter. Read all parts to each question before you begin.

You have learned energy cannot be created nor destroyed, but that it can change form. Explain how hydroelectric energy can be used to make electricity. Include details about the changes of form the energy goes through in order for electricity to be produced.

HINT Use the text and illustrations on page F104 to answer this question.

_____

_____

_____

_____

_____

_____

If you have ever been to the ocean, you have probably watched the rise and fall of the water known as tides. In certain areas, the energy of the tides can be harnessed to produce electricity. Explain how this process works, and why tidal energy isn't used along all coastlines.

HINT Reread the text on page F106 and F107 to help you answer this question.

_____

_____

_____

_____

_____

_____

_____

Sources of Energy

Fossil fuels and hydroelectric energy are the major energy sources throughout the world. Fossil fuels are the main source of energy for industrial nations like the United States. Two alternative energy sources are nuclear and geothermal energy.

Write a letter to the editor in which you compare and contrast nuclear energy and geothermal energy. Include in your letter information on how each type of energy source is harnessed and used.

Use this page for prewriting or planning activities. Then write your response on a separate sheet of paper.

Writer's Checklist

IDEAS

- Is my message clear?
- Do I know enough about my topic?
- Have I included interesting details?

ORGANIZATION

- Does my paper start out with a bang?
- Did I tell things in the best order?
- At the end does it feel finished and make you think?

VOICE

- Does this writing really sound like me?
- Did I say what I was thinking?
- Did I express how I feel?

WORD CHOICE

- Will my reader understand my words?
- Did I use words I love?
- Are my words interesting?
- Can I picture it?

SENTENCE FLUENCY

- Is my paper easy to read out loud?
- Do my sentences begin in different ways?
- Are some sentences long and some short?

CONVENTIONS

- Did I use paragraphs?
- Is it easy to read my spelling?
- Did I use capital letters in the right place?
- Are periods, commas, exclamation marks, and quotation marks in the right places?

Math Practice

Unit F, Chapter 1

Read each question and choose the best answer. Mark the letter for that answer.

Use the table below for problems 1 and 2.

Weight of 100-Pound Person on Different Planets

Planet	Weight (pounds)
Earth	100
Jupiter	264
Mars	38
Pluto	0.6

1. Which bar graph best represents the data in the table?

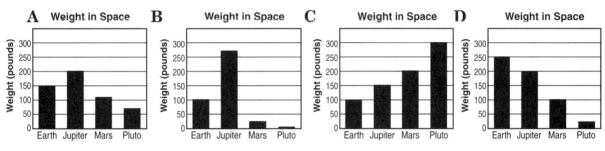

2. Determine the difference between the weight of the person on Earth compared to that on Pluto.

  A 100.6 lbs.   **C** 99.4 lbs.

  B 100 lbs.   **D** 98.4 lbs.

3. Suppose two people are moving a large box and pushing with a force of 140 newtons. The opposing force of friction is 49 newtons. What is the net force?

  A 189 N   **C** 100 N

  B 140 N   **D** 91 N

4. How many joules of work does a 342-newton student do to climb a 5.7-meter flight of stairs?

  A 1949.4 J   **C** 336.3 J

  B 347.7 J   **D** 300 J

Use the table below for problems 5 and 6.

Work Needed to Move an Object 4 Meters

Weight of Object	Work
4 N	16 J
10 N	40 J
30.2 N	120.8 J
50.7 N	202.8 J

5. What type of graph would best represent the data in the table?

  A circle graph

  B line graph

  C stem-and-leaf plot

  D pictograph

6. How much more work was required to move the 50.7 N object than the 4 N object?

  A 46.7 J   **C** 186.8 J

  B 82 J   **D** 202.8 J

Unit F, Chapter 2

Read each question and choose the best answer. Mark the letter for that answer.

7. Speed equals distance divided by time. What is the speed of a scooter that travels 40 miles in 2.5 hours?

   A 16 miles per hour

   B 40 miles per hour

   C 42.5 miles per hour

   D 100 miles per hour

8. If two objects have the same velocity, the one with the greater mass has more momentum. Choose the statement that is TRUE.

   A A large car has less momentum than a small car moving at the same speed.

   B A large car has more momentum than a small car moving at the same speed.

   C A large car has the same momentum as a small car moving at the same speed.

   D Moving cars do not have momentum.

9. For an experiment, you are asked to cut a string that is 243 centimeters long. How many meters is this?

   A 243 m

   B 24.3 m

   C 2.43 m

   D 0.243 m

10. What is the best unit to use to measure the distance a ball rolls?

   A liters

   B grams

   C meters

   D newtons

11. What is the best unit to use to measure the mass of a small ball?

   A liters

   B grams

   C meters

   D newtons

12. What is the best unit of measure to use to time how long it takes a basketball to stop rolling on a level surface?

   A minutes

   B hours

   C days

   D years

Unit F, Chapter 3

Read each question and choose the best answer. Mark the letter for that answer.

Use the table below for problems 13 and 14.

Tennis Ball Drop

Drop Height	Bounce Height
50 cm	40.5 cm
100 cm	80.0 cm
150 cm	121.5 cm

13. Choose the line graph that best represents the data in the table.

A

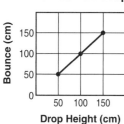

C

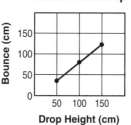

B

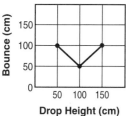

D

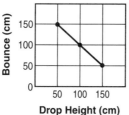

14. How much higher is the bounce of a ball dropped from 150 centimeters than a ball dropped from 50 centimeters?

A 81 cm

B 80 cm

C 40.5 cm

D 8 cm

15. A 40-watt bulb uses 40 watts of power every second. How many watts would be used in a quarter of an hour if the bulb was not turned off?

A 36,400 watts   **C** 600 watts

B 36,000 watts   **D** 900 watts

16. Angle 1 measures 120°. Angle 2 measures 37°. What is the difference in the angles?

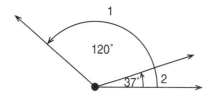

A 157°   **C** 83°

B 120°   **D** 37°

Use the table below for problems 17 and 18.

**Speed of Sound Waves
Through Different Materials**

Material	Speed (meters per second)
Air	340
Water	150
Steel	5000
Silver	2650
Granite	3950

17. What is the median of the data in the table?

A 5000   **C** 2418

B 2650   **D** 150

18. What is the mean of the data in the table?

A 5000   **C** 2418

B 2650   **D** 150

Unit F, Chapter 4

Read each question and choose the best answer. Mark the letter for that answer.

19. One calorie is the amount of heat needed to raise the temperature of 1 gram of water by 1°C. How many calories would be needed to raise the temperature of 5.7 grams of water from 0°C to 100°C?

   A 5.7 calories

   B 57 calories

   C 570 calories

   D 5700 calories

20. Tides in the Bay of Fundy can rise and fall more than 15 meters. How many centimeters is this?

   A 15,000 cm

   B 1500 cm

   C 150 cm

   D 15 cm

21. During an activity you are asked to cut 150 strips of paper, each 3.5 centimeters long. How many meters of paper will you need?

   A 525 m

   B 52.5 m

   C 50.25 m

   D 5.25 m

22. The blades of some turbines are 330 feet across. How many yards is this?

   A 990 yd

   B 110 yd

   C 11 yd

   D 3 yd

Use the circle graph below for problems 23 and 24.

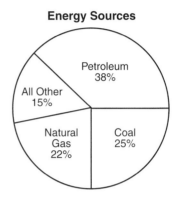

Energy Sources

23. Determine the most common source of energy.

   A natural gas

   B coal

   C petroleum

   D other sources

24. Determine what percent more of the current energy sources is petroleum than natural gas.

   A 23 %

   B 16 %

   C 13 %

   D 10 %

Unit F, Review

Read each question and choose the best answer. Mark the letter for that answer.

25. A person is moving a small box and pushing with a force of 40 newtons. The opposing force of friction from the box is 23.7 newtons. What is the net force?

- **A** 40 newtons
- **B** 23.7 newtons
- **C** 16.3 newtons
- **D** 6.3 newtons

26. How many joules of work does a 420-newton person do to climb a 3.8-meter flight of stairs?

- **A** 1596 J
- **B** 1423 J
- **C** 423.8 J
- **D** 416.2 J

27. What is the best unit to use to measure the volume of a liquid?

- **A** meters
- **C** newtons
- **B** joules
- **D** liters

28. One calorie is the amount of heat needed to raise the temperature of one gram of water by one degree Celsius. How many calories would be needed to raise the temperature of 3.8 grams of water from 20°C to 130°C?

- **A** 110 calories
- **B** 130 calories
- **C** 200 calories
- **D** 418 calories

Use the table below for problems 29 and 30.

Speed of Sound Waves Through Different Materials

Material	Speed (meters per second)
Silver	2650
Steel	5000
Granite	3950

29. Choose the bar graph that best represents the data in the table.

A
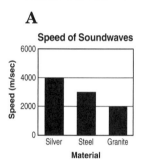

C
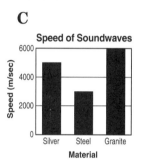

B
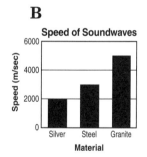

D
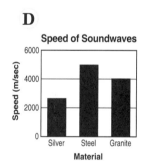

30. Find the mean for the data in the table. Round your answer to the nearest whole number.

- **A** 2000 meters per second
- **B** 2650 meters per second
- **C** 3867 meters per second
- **D** 3950 meters per second